Instructor's Solutions Manual for

Yates, Moore, and Starnes's

The Practice of Statistics

TI-83/89 Graphing Calculator Enhanced

Second Edition

Christopher Barat

Virginia State University

W. H. Freeman and Company
New York

Printed in the United States of America

ISBN: 0-7167-8344-4 (EAN: 9780716783442)

Third printing

Contents

Exploring Data

1.1 (a) The individuals are vehicles (or "cars"). (b) The variables are: vehicle type (categorical), transmission type (categorical), number of cylinders (quantitative), city MPG (quantitative), and highway MPG (quantitative).

1.2 (a) Categorical. (b) Quantitative. (c) Categorical. (d) Categorical. (e) Quantitative. (f) Quantitative.

1.3 Possible answers (units):
- Number of pages (pages)
- Number of chapters (chapters)
- Number of words (words)
- Weight or mass (pounds, ounces, kilograms . . .)
- Height and/or width and/or thickness (inches, centimeters . . .)
- Volume (cubic inches, cubic centimeters . . .)

1.4 Possible answers (reasons should be given): unemployment rate, average (mean or median) income, quality/availability of public transportation, number of entertainment and cultural events, housing costs, crime statistics, population, population density, number of automobiles, various measures of air quality, commuting times (or other measures of traffic), parking availability, taxes, quality of schools.

1.5 (a) Shown below. The bars are given in the same order as the data in the table—the most obvious way—but that is not necessary (since the variable is nominal, not ordinal). (b) A pie chart would not be appropriate, since the different entries in the table do not represent parts of a single whole.

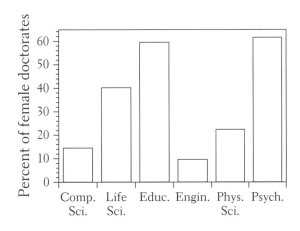

1.6 (a) Below. For example, "Motor Vehicles" is 46% since $\frac{42{,}340}{92{,}353} = 0.45846\ldots$. The "Other causes" category is needed so that the total is 100%. (b) Below. The bars may be in any order. (c) A pie chart *could* also be used, since the categories represent parts of a whole (all accidental deaths).

Cause	Percent
Motor vehicles	46
Falls	13
Drowning	4
Fires	4
Poisoning	11
Other causes	22

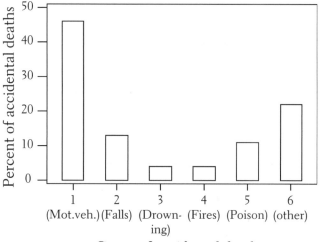

1.7

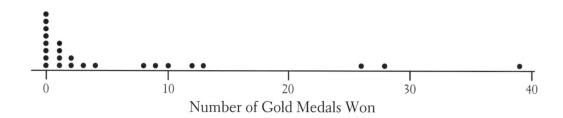

The distribution has a peak at 0 and a long right tail. There are eight outliers, with the most severe being 26, 28, and 39. The spread is 0 to 39 and the center is 1.

1.8

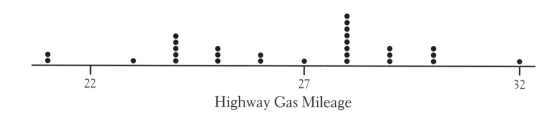

The distribution is skewed to the left, with a major peak at 28 and a minor peak at 24. The spread is relatively narrow (21 to 32 mpg). The two observations at 21 and the observation at 32 appear to be mild outliers. The center is 28 mpg.

1.9 (a) Stems = thousands, leaves = hundreds. The data have been rounded to the nearest $100.
(b) The distribution is skewed strongly to the right, with a peak at the 1 stem. The spread is approximately 19,000 ($1300 to $19,300). The center is 45 ($\approx$ $4500). The observations 182 ($\approx$ $18,200) and 193 ($\approx$ $19,300) appear to be outliers.

1.10

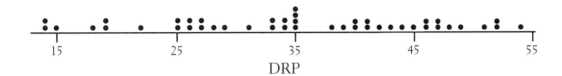

DRP

The center of the distribution is 35, and there are approximately the same number of points to the left and right of the center. There are no major gaps or outliers. The distribution is approximately symmetric.

1.11 (a)

```
0 | 399
1 | 1345677889
2 | 000123455668888
3 | 25699
4 | 1345579
5 | 0359
6 | 1
7 | 0
8 | 366
9 | 3
```

(b)

```
0 | 3
0 | 99
1 | 134
1 | 5677889
2 | 0001234
2 | 55668888
3 | 2
3 | 5699
4 | 134
4 | 5579
5 | 03
5 | 59
6 | 1
6 |
7 | 0
7 |
8 | 3
8 | 66
9 | 3
```

Both plots show the general shape of the distribution; however, the split-stem plot may be preferable since it shows more detail.

(c) The distribution is skewed to the right with a peak in the 2 stem(s). The spread is approximately 90 (3 to 93). There are several moderate outliers visible in the split-stem plot; specifically, the five amounts of $70 or more. While most shoppers spent small to moderate amounts of money, a "cluster" of shoppers spent larger amounts ranging from $70 to $93. The center of the distribution is at approximately $28.

1.12 (a)

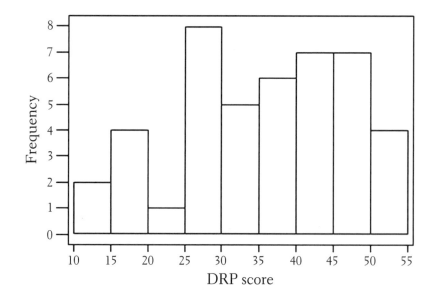

Percent	Freq.
5.0–5.9	1
6.0–6.9	0
7.0–7.9	0
8.0–8.9	1
9.0–9.9	1
10.0–10.9	2
11.0–11.9	9
12.0–12.9	11
13.0–13.9	14
14.0–14.9	6
15.0–15.9	4
16.0–16.9	0
17.0–17.9	0
18.0–18.9	1
Total	50

(b) The distribution is slightly skewed to the left with a peak at the class 13.0–13.9. There is one outlier in each tail of the distribution.

1.13

DRP Score	Freq.
10–14	2
15–19	4
20–24	1
25–29	8
30–34	5
35–39	6
40–44	7
45–49	7
50–54	4
Total	44

The dotplot provides more detail, but the histogram has the advantage of clearly displaying two "clusters" of DRP scores (the classes 25–29 and 40–44, 45–49).

1.14

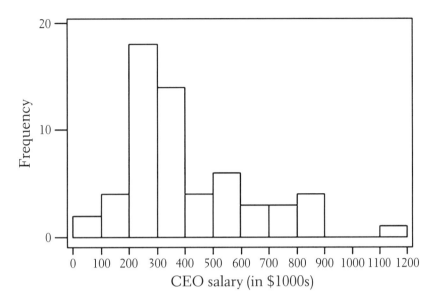

The distribution is skewed to the right with a peak in the 200s class. The spread is approximately 1100 ($21,000 to $1,103,000) and the center is located at 350 ($350,000). There is one outlier in the 1100s class.

1.15 (b) The distribution is symmetric with a peak at class (chest size) 40. The center is also located at 40. The spread is 15 (33 to 48). Assuming that the sample is representative of all members of the population, the distribution would provide a useful guide to those making clothing for the militiamen. From the frequency table, it is easy to estimate the percentage of all militiamen who have a certain chest size. The production of uniforms can reflect this distribution.

1.16 (a) Roughly symmetric, though it might be viewed as SLIGHTLY skewed to the right. (b) About 15%. (39% of the stocks had a total return less than 10%, while 60% had a return less than 20%. This places the center of the distribution somewhere between 10% and 20%.) (c) The smallest return was between −70% and −60%, while the largest was between 100% and 110%. (d) 23% (1 + 1 + 1 + 1 + 3 + 5 + 11).

1.17 (a) Skewed to the right; center at about 3 (31 less than 3, 11 equal to 3, 23 more than 3); spread: 0 to 10. No outliers. (b) About 23% (15 out of 65 years).

1.18 Lightning histogram: centered at noon (or more accurately, somewhere from 11:30 to 12:30). Spread is from 7 to 17 (or more accurately, 6:30 AM to 17:30, i.e., 5:30 PM). Shakespeare histogram: centered at 4, spread from 1 to 12.

1.19 (a)

Percent	Cumulative frequency	Relative cumulative frequency	Percent	Cumulative frequency	Relative cumulative frequency
5.0–5.9	1	2%	12.0–12.9	25	50%
6.0–6.9	1	2%	13.0–13.9	39	78%
7.0–7.9	1	2%	14.0–14.9	45	90%
8.0–8.9	2	4%	15.0–15.9	49	98%
9.0–9.9	3	6%	16.0–16.9	49	98%
10.0–10.9	5	10%	17.0–17.9	49	98%
11.0–11.9	14	28%	18.0–18.9	50	100%

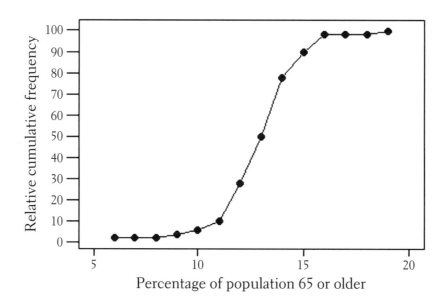

(b) • Percentage of states in which percentage of "65 and older" is less than 15% = 90%, since the point (15, 90) lies on the ogive.

   • 40th percentile of distribution ≈ 12.4%, since the horizontal line drawn from 40% on the vertical axis intersects the ogive at a point whose horizontal coordinate is approximately 12.4%. Less than 40% of states have 12.4% or less of their population aged 65 or older.

   • Answers vary.

1.20 (a) The center corresponds to the 50th percentile. Draw a horizontal line from the value 50 on the vertical axis and determine the point on the ogive where the line intersects the ogive. Then draw a vertical line from this point to the horizontal axis. The line intersects the axis at approximately $28. Thus, $28 is the estimate of the center.
(b) The 20th percentile.
(c)

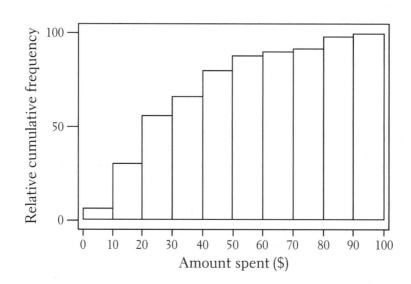

1.21 (a)

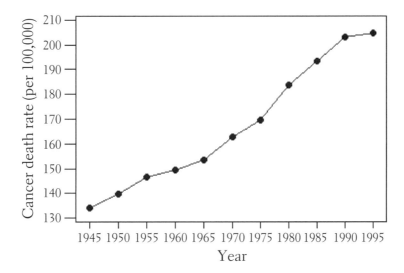

The cancer death rate has risen steadily from 1945 to 1995, with the largest increase occurring in the period 1975–1980.

(b) No, the slower rate of increase during the period 1990–1995 suggests that some progress was made during that time (at least in terms of treating the disease effectively). However, we have yet to see a decrease in the death rate, indicating that much work remains to be done in terms of actively preventing the disease.

1.22 (a)

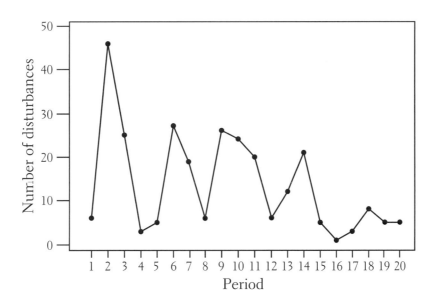

[Note: the periods are numbered consecutively from period 1, Jan.–Mar. 1968, to period 20, Oct.–Dec. 1972, on the horizontal axis.]

(b) The plot shows a decreasing trend—fewer disturbances overall in the later years—and more importantly, there is an apparent cyclic behavior. Looking at the table, the spring and summer months (April through September) generally have the most disturbances—probably for the simple reason that more people are outside during those periods.

1.23 Gender, party voted for: Categorical
        Age, income: Quantitative

1.24 (a) Car makes: a bar chart or pie chart. Car age: a histogram or stemplot. (b) Study time: a histogram or stemplot. Change in study hours: a time plot (average hours studied vs. time). (c) A bar chart or pie chart.

1.25 An "Other Methods" plot is needed because the sum of the percentages for the other categories is less than 100%.

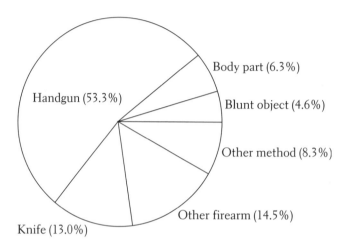

1.26 (a)

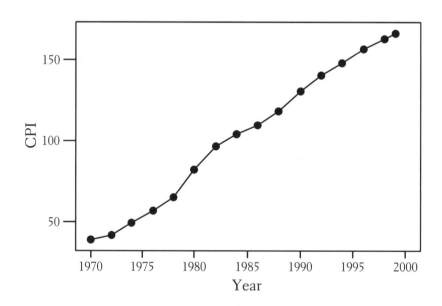

(c) Prices rose steadily during this period. There was no reversal of this trend in any of the periods under study.

(d) Prices were rising fastest during the mid- to late 1970s and rising slowest during the early 1970s and the mid-1980s.

1.27 (a)

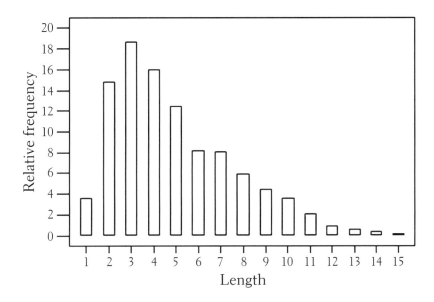

The distribution is skewed to the right with a single peak. There are no gaps or outliers.

(b) Shakespeare was somewhat more likely to use short words and somewhat less likely to use extremely long words than *Popular Science*. However, the distributions have strongly similar shapes.

1.28

```
48 | 8
49 |
50 | 7
51 | 0
52 | 6799
53 | 04469
54 | 2467
55 | 03578
56 | 12358
57 | 59
58 | 5
```

Stem = first two digits   Leaf = last digit.

The distribution is roughly symmetric with one value (4.88) that is somewhat low. The center of the distribution is between 5.4 and 5.5.

Based on the plot, we would estimate the Earth's density to be about halfway between 5.4 and 5.5.

1.29 (a)

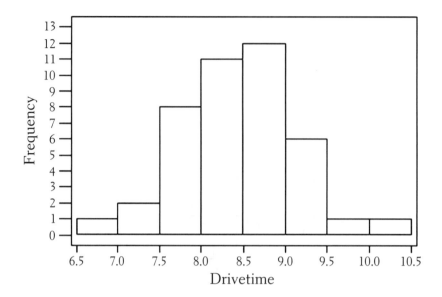

The distribution is roughly symmetric with no clear outliers.

(b)

Drivetime	Cum. freq.	Rel. cum. freq.
7.0	1	2.4%
7.5	3	7.1%
8.0	12	28.6%
8.5	23	54.8%
9.0	35	83.3%
9.5	40	95.2%
10.0	41	97.6%
10.5	42	100%

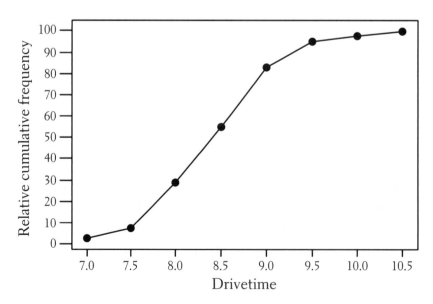

(c) Center ≈ 8.5, 90th percentile ≈ 9.4

(d) 8.0 ≈ 28th percentile

1.30 (a)

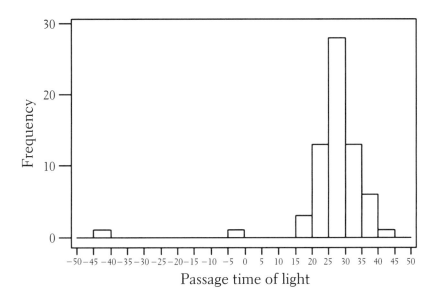

A stemplot would have much the same appearance as the histogram, but it would be somewhat less practical, because of the large number of observations with common stems (in particular, the stems 2 and 3).

(b) The histogram is approximately symmetric with two unusually low observations at −44 and −2. Since these observations are strongly at odds with the general pattern, it is highly likely that they represent observational errors.

(c)

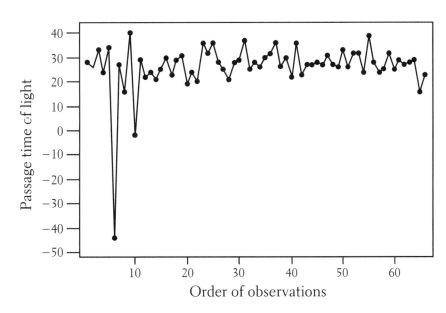

(d) Newcomb's worst measurement errors occurred early in the observation process. As the observations progressed, they became more consistent.

1.31 (a) $n = 14$, $\Sigma x = 1190$. The mean is $\bar{x} = \dfrac{\Sigma x}{n} = \dfrac{1190}{14} = 85$.

(b) If the 15th score is 0, then $n = 15$, $\Sigma x = 1190$, and the new mean is

$$\bar{x} = \frac{\Sigma x}{n} = \frac{1190}{15} = 79.3$$

The fact that this value of $\bar{x}$ is less than 85 indicates the nonresistance property of $\bar{x}$. The extremely low outlier at 0 pulled the mean below 85.

(c) Minitab splits the decades to show greater detail.

```
Stem-and-leaf of C1  N = 14
Leaf Unit = 1.0
7   4
7   568
8   024
8   67
9   013
9   68
```

And here is a histogram, with the widths of the bars specified to correspond to letter grades: D (68–75), C (76–83), B (84–91), and A (92–100). Both plots show a fairly balanced or symmetric distribution, with the histogram suggesting a slight skewness to the left. (Note that the mean and the median are the same (85).)

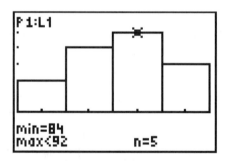

Given a rather small data set like this one, the stem plot would normally be preferable. But since we are very interested in letter grades in this case, perhaps the histogram would be most informative.

1.32 (a)

```
10 | 139
11 | 5
12 | 669
13 | 77
14 | 08
15 | 244
16 | 55
17 | 8
18 |
19 |
20 | 0
```

200 is a potential outlier. The center is approximately 140. The spread (excluding 200) is 77.

(b) $\bar{x} = 2539/18 = 141.058$.

(c) Median = average of ninth and tenth scores = 138.5. The mean is larger than the median because of the outlier at 200, which pulls the mean towards the long right tail of the distribution.

1.33 Since the mean $\bar{x} = \$1.2$ million and the number of players on the team is $n \bullet 25$, the team's annual payroll is

$$(\$1.2 \text{ million}) (25) = \$30 \text{ million}$$

If you knew only the median salary, you would not be able to calculate the total payroll because you cannot determine the *sum* of all 25 values from the median. You can only do so when the arithmetic average of the values is provided.

1.34 $\bar{x} = \frac{\$480,000}{8} = \$60,000$. Seven of the eight employees (everyone but the owner) earned less than the mean. The median is $M = \$22,000$.

A recruiter might try to mislead applicants by telling them the mean salary, rather than the median salary, when the applicants ask about the "average" or "typical" salary. The median is a far more accurate depiction of a "typical" employee's earnings, because it resists the effects of the outlier at $270,000.

1.35 Mean = $675,000; median = $330,000. The mean is nonresistant to the effects of the extremely high incomes in the right tail of the distribution. It will therefore be larger than the median.

1.36 (a)

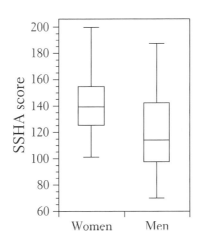

(b)

	$\bar{x}$	M	Five-number summaries				
Women	141.06	138.5	101	126	138.5	154	200
Men	121.25	114.5	70	98	114.5	143	187

(c) All the displays and descriptions reveal that women generally score higher than men. The men's scores ($IQR = 45$) are more spread out than the women's (even if we don't ignore the outlier). The shapes of the distributions are reasonably similar, with each displaying right skewness.

1.37 (a) The mean and median should be approximately equal since the distribution is roughly symmetric.

(b) Five-number summary: 42, 51, 55, 58, 69 $\quad \bar{x} = 2357/43 = 54.8$

As expected, median and $\bar{x}$ are very similar.

(c) Between Q_1 and Q_3: 51 to 58.

(e)

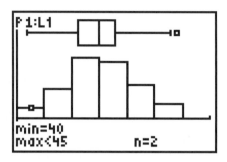

The point 69 is an outlier; this is Ronald Reagan's age on inauguration day. W. H. Harrison was 68, but that is not an outlier according to the 1.5(IQR) test.

1.38 Yes, IQR is resistant. Take the data set 1, 2, 3, 4, 5, 6, 7, 8 as an example. In this case the median = 4.5, $Q_1 = 2.5$, $Q_3 = 6.5$, and IQR = 4. Changing any "extreme" value (that is, any value outside the interval between Q_1 and Q_3) will have no effect on the IQR. For example, if 8 is changed to 88, IQR will still equal 4.

1.39 (a) $\bar{x} = 34.7022$ and $n = 50$: Thus, total amt. spent = (34.7022)(50) = \$1735.11.

(b)

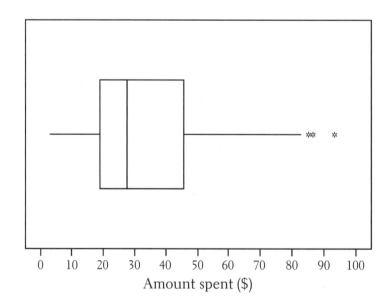

Amount spent (\$)

The boxplot indicates the presence of several outliers. According to the 1.5 × IQR rule, these outliers are 85.76, 86.37, and 93.34.

1.40 (a) $\bar{x} = 32.4 \div 6 = 5.4$. (b) $\sum(x_i - \bar{x})^2 = (0.2)^2 + (-0.2)^2 + (-0.8)^2 + (-0.5)^2 + (0.3)^2 + (1.0)^2 = 2.06$; $s^2 = 2.06 \div 5 = 0.412$; $s = \sqrt{0.412} = 0.6419$. (c) Yes, they agree: $\bar{x} = 5.4$, $s = 0.6418722614 \approx 0.6419$.

1.41 (a) $\sum(x_i - \bar{x})^2 = (-12.1)^2 + (1.9)^2 + (-10.1)^2 + (12.9)^2 + (34.9)^2 + (6.9)^2 + (-3.1)^2 + (-.1)^2 + (-18.1)^2 + (-13.1)^2 = 2192.9.$
$s^2 = 2192.9/9 = 243.66, s = 15.609.$

(b) Excluding the outlier at 61, we obtain $\bar{x} = 22.2, s = 10.244$. The outlier caused the values of both measures to increase; the increase in s is more substantial. Clearly, s is not a resistant measure of spread.

1.42 The five-number summary is somewhat preferable due to the skewness of the distribution.
Five-number summary: 5.5, 11.5, 12.75, 13.725, 18.3.

1.43 (a) 1, 1, 1, 1. (b) 0, 0, 10, 10. (c) For (a), any set of four identical numbers will have $s = 0$. For (b), the answer is unique; here is a rough description of why. We want to maximize the "spread-out"-ness of the numbers (that is what standard deviation measures), so 0 and 10 seem to be reasonable choices based on that idea. We also want to make each individual squared deviation— $(x_1 - \bar{x})^2, (x_2 - \bar{x})^2, (x_3 - \bar{x})^2,$ and $(x_4 - \bar{x})^2$—as large as possible. If we choose 0, 10, 10, 10—or 10, 0, 0, 0—we make the first squared deviation (7.5^2), but the other three are only $(2.5)^2$. Our best choice is two at each extreme, which makes all four squared deviations equal to 5^2.

1.44 (a) $\bar{x} = 7.5/5 = 1.5, s = .436.$

(b) To obtain $\bar{x}$ and s in centimeters, multiply the results in inches by 2.54: $\bar{x} = 3.81$ cm, $s = 1.107$ cm.

(c) The average cockroach length can be estimated as the mean length of the five sampled cockroaches: that is, 1.5 inches. This is, however, a questionable estimate, because the sample is so small.

1.45 (a) The mean and the median will both increase by $1000.

(b) No. Each quartile will increase by $1000, thus the difference $Q_3 - Q_1$ will remain the same.

(c) No. The standard deviation remains unchanged when the same amount is added to each data value.

1.46 A 5% across-the-board raise will cause both IQR and s to increase. The transformation being applied here is $x^* = 1.05x$, where $x =$ the old salary and $x^* =$ the new salary. Both IQR and s will increase by a factor of 1.05.

1.47 Calories: There seems to be little difference between beef and meat hot dogs, but poultry hot dogs are generally lower in calories than the other two. In particular, the median number of calories in a poultry hot dog is smaller than the lower quartiles of the other two, and the poultry lower quartile is less than the minimum calories for beef and meat.

Type	Min	Q_1	M	Q_3	Max
Beef	111	140	152.5	178.5	190
Meat	107	138.5	153	180.5	195
Poultry	86	100.5	129	143.5	170

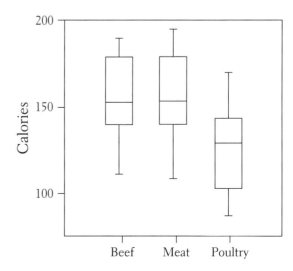

Sodium: Overall, beef hot dogs have less sodium (except for the one with the most sodium: 645 mg). Even if we ignore the low outlier among meat hot dogs, meat holds a slight edge over poultry.

Type	Min	Q_1	M	Q_3	Max
Beef	253	320.5	380.5	478	645
Meat	144	379	405	501	545
Poultry	357	379	430	535	588

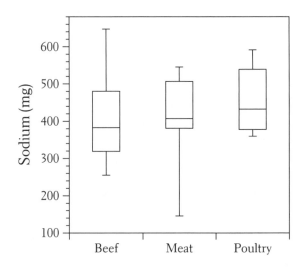

1.48 (a)

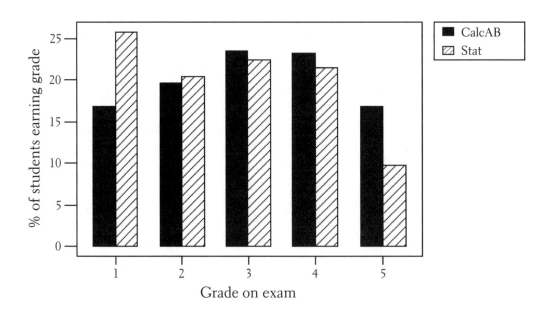

(b) The two distributions are roughly similar for grades of 2, 3, and 4. The major differences occur for grades 1 and 5. A considerably larger proportion of Statistics students receive a grade of 1, and a considerably smaller proportion of Statistics students receive a grade of 5. This suggests that the Statistics exam is harder, at least in the sense that students are more likely to get very poor grades on the Statistics exam than on the Calculus AB exam.

1.49 (a)

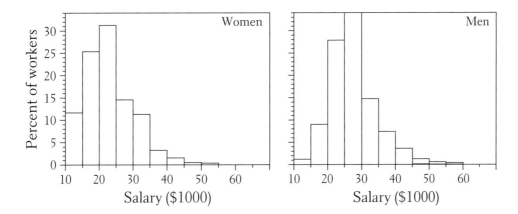

Use relative frequency histograms, since there are considerably more men than women.

(b) The two histograms are both skewed to the right (as income distributions often are). Women's salaries are generally lower than men's.

(c) The women's total sums to 100.1%, due to roundoff error.

1.50 (a)

V-AA			I-AAA
776	3		8
44	4		1
65	4		677
4	5		1234
8	5		568
20	6		2233444
97766	6		5578
42	7		114
	7		678
	8		
6	8		7
32	9		1
886	9		
	10		
	10		6

(b) The I-AAA point distribution is skewed to the right, while the V-AA distribution is roughly symmetric. The median of the I-AAA distribution is 63.5, and the median of the V-AA distribution is 66; they are roughly equivalent. The I-AAA distribution is slightly more spread out, due in large part to the outlier at 106. The median and quartiles are the best measures for comparison purposes, because one of the distributions is skewed.

V-AA		I-AAA
1	0	222
8	0	5578
4	1	1444
	1	7
300	2	
55	2	
1	3	03
	3	
2	4	0
7	4	
3	5	
	5	
	6	0

A back-to-back stemplot of the margin of victory distributions reveals that large margins of victory were more likely to occur in the V-AA games than in the I-AAA games (despite the few I-AAA outliers). The medians of the V-AA and I-AAA distributions are 24 and 12.5, respectively, which further suggests that the average margin of victory tended to be larger in the V-AA games.

1.51

10	7
11	
12	
13	5689
14	067
15	3
16	
17	2359
18	2
19	015

There are two distinct clusters of brands (the distribution has two peaks with a gap between them) and one outlier in the lower tail. The boxplot hid the clusters. The quartiles are roughly at the centers of the two clusters, so much of the IQR is the spread between the clusters. Because of this, the $1.5 \times IQR$ rule used in drawing the boxplot did not call attention to the outlier.

1.52

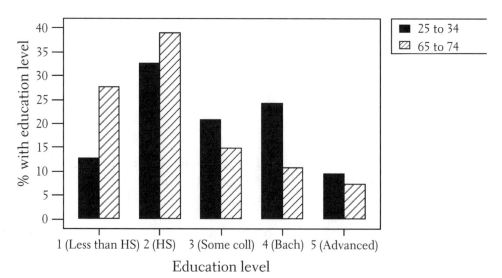

The side-by-side bar graph shows several distinct differences in educational attainment between the two groups. The 65-to-74 age group was more likely to have earned no more than a high school diploma. The 25-to-34 age group was more likely to have gone to college and to have completed a Bachelor's degree. However, the percentages in the "Advanced" group are relatively similar, indicating that those 65-to-74 year-olds who managed to complete college were more likely to earn an advanced degree.

1.53

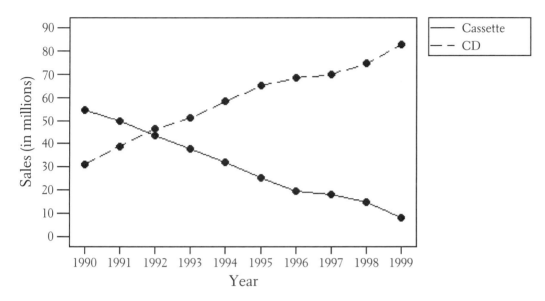

Over the period 1990–1999, sales of cassettes declined steadily, while sales of CD's increased steadily. The first year during which sales of CD's exceeded sales of cassettes was 1992.

1.54 The means and standard deviations are basically the same. For Set A, $\bar{x} \approx 7.501$ and $s \approx 2.032$, while for Set B, $\bar{x} \approx 7.501$ and $s \approx 2.031$. Set A is skewed to the left, while Set B has a high outlier.

	Set A			Set B
3	1		5	257
4	7		6	58
5			7	079
6	1		8	48
7	2		9	
8	1177		10	
9	112		11	
			12	5

1.55 (a) $x^* = 746x$, where x = measurement in horsepower and x^* = measurement in watts. The mean, median, IQR, and standard deviation will all be multiplied by 746.

(b) $x^* = (5/9)(x - 32)$, where x = measurement in °F, x^* = measurement in °C. The mean and median will be multiplied by 5/9 and the result reduced by $(5/9)(32) = 160/9$. The IQR and standard deviation will be multiplied by 5/9.

(c) $x^* = x + 10$, where x = original test score, x^* = "curved" test score. The mean and median will increase by 10. The IQR and standard deviation will remain the same.

1.56 Variance is changed by a factor of $2.54^2 = 6.4516$; generally, for a transformation $x_{\text{new}} = a + bx$, the new variance is b^2 times the old variance.

1.57 (a) Five-number summary for normal corn: 272, 337, 358, 400.5, 462. Five-number summary for new corn: 318, 383.5, 406.5, 428.5, 477. The boxplots show that the new corn seems to increase weight gain—in particular, the median weight gain for new-corn chicks was greater than Q_3 for those that ate normal corn.

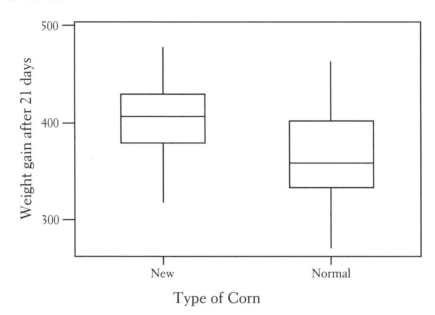

(b) Normal corn: $\bar{x} = 366.3$, $s = 50.805$; new corn: $\bar{x} = 402.95$, $s = 42.729$. On the average, the chicks that were fed the new corn gained 36.65 grams more mass (weight) than the other chicks.

(c) Means and standard deviations will all be multiplied by 1/28.35 in order to convert grams to ounces. Normal corn: $\bar{x} = 12.92$, $s = 1.792$; new corn: $\bar{x} = 14.21$, $s = 1.507$.

1.58 (a) Mean—although incomes are likely to be right-skewed, the city government wants to know about the total tax base. (b) Median—the sociologist is interested in a "typical" family, and wants to lessen the impact of the extremes.

1.59 Possible answers are total profits, number of employees, total value of stock, and total assets.

1.60 (a) All major league baseball players as of opening day, 1998.
(b) Team (categorical), position (categorical), age (quantitative), salary (quantitative).
(c) Age is measured in years, salary in $1000s (thousands of dollars).

1.61 (a)

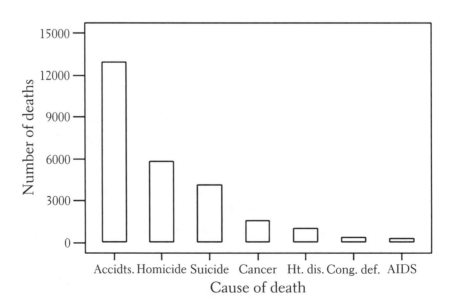

(b) To make a pie chart, you need to know the number of people in the "death from other causes" category. Including "death from other causes" ensures that the category percentages will sum to 100%.

1.62 (a) Since a person cannot choose the day on which he or she has a heart attack, one would expect that all days are "equally likely"—no day is favored over any other. While there is *some* day-to-day variation, this does seem to be supported by the chart.

(b) Monday through Thursday are fairly similar, but there is a pronounced peak on Friday, and lows on Saturday and Sunday. Patients do have some choice about when they leave the hospital, and many probably choose to leave on Friday, perhaps so that they can spend the weekend with the family. Additionally, many hospitals cut back on staffing over the weekend, and they may wish to discharge any patients who are ready to leave before then.

1.63 (a)

```
4 | 33
4 |
4 |
4 | 89
5 | 0001
5 |
5 | 45
5 | 7
5 | 9
6 | 11
```

(b) Median = 50.4.

(c) Q_3 = 57.4. Landslides occurred in 1956, 1964, 1972, and 1984.

1.64 Slightly skewed to the right, centered at 4.

1.65 (a)

MPG of Year 2000 SUVs

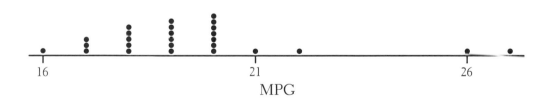

MPG

The distribution is skewed to the right with a peak at 20. There are two outliers, at 26 and 27 (Toyota RAV4 and Subaru Forester, respectively).

Variable	N	Mean	Median	StDev
MPG	26	19.500	19.000	2.470

Variable	Minimum	Maximum	Q1	Q3
MPG	16.000	27.000	18.000	20.000

The fact that $\bar{x}$ > M reflects the right skewness of the distribution.

(b)

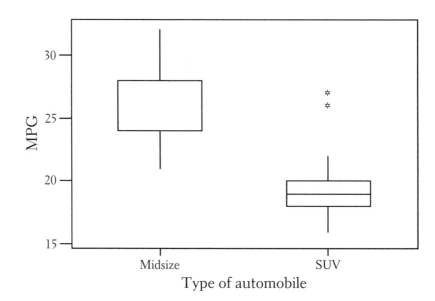

The boxplots clearly show that the midsize automobiles tend to have higher MPGs than the SUVs. The only SUV observations falling in the "middle" of the midsize distribution are the two outliers at 26 and 27.

(Note: the lack of a "median line" in the midsize boxplot is the result of the median and Q_3 being the same value.)

1.66 (a)

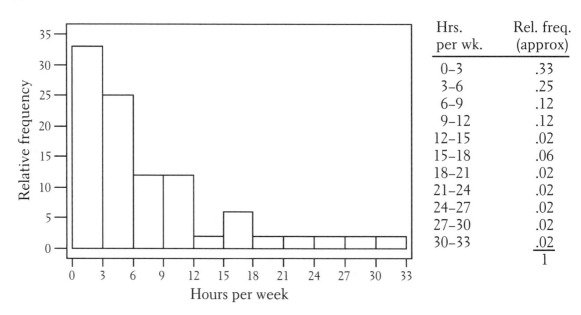

Hrs. per wk.	Rel. freq. (approx)
0–3	.33
3–6	.25
6–9	.12
9–12	.12
12–15	.02
15–18	.06
18–21	.02
21–24	.02
24–27	.02
27–30	.02
30–33	.02
	1

(b) From the ogive: Median $\approx$ 6 (50th percentile), $Q_1 \approx$ 2.5 (25th percentile), $Q_3 \approx$ 11 (75th percentile). There are outliers, according to the $1.5 \times$ IQR rule, because values exceeding $Q_3 + (1.5 \times$ IQR$) = 23.75$ clearly exist.

(c) 10 hours $\approx$ 70th percentile.

1.67 (a) Min = −34.04, Q_1 = −2.95, Med = 3.47, Q_3 = 8.45, Max = 58.68.

(b) The distribution is fairly symmetric, with a single peak in the high single digits (5 to 9). There are no gaps, but four "low" outliers and five "high" outliers are listed separately.

(c) 58.68% of $1000 is $586.60. The stock is worth $1586.50 at the end of the best month. In the worst month, the stock lost 1000(.3404) = $340.40, so the $1000 decreased in worth to 1000 − 340.40 = $659.60.

(d) IQR = $Q_3 − Q_1$ = 8.45 − (−2.95) = 11.4

$1.5 \times$ IQR = 17.1

$Q_1 − (1.5 \times$ IQR) = −2.95 − 17.1 = −20.05

$Q_3 + (1.5 \times$ IQR) = 8.45 + 17.1 = 25.55

The four "low" and five "high" values are all outliers according to the criterion. It does appear that SPLUS uses the $1.5 \times$ IQR criterion to identify outliers.

1.68 The difference in the mean and median indicates that the distribution of awards is skewed sharply to the right—that is, there are some *very* large awards.

1.69 The median—half are traveling faster than you, and half are traveling slower. (Actually, you have found *a* median—it could be that a whole range of speeds, say from 56 mph to 58 mph, might satisfy this condition.)

1.70

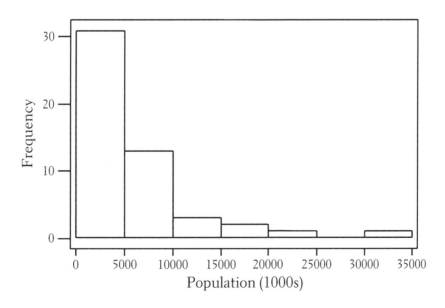

The distribution is strongly skewed right, with a spread of approximately 33,000 and a center located in the first (0–5000) class. The distribution is not surprising inasmuch as the many small to medium-sized U.S. states have small populations and a few states, in particular California, Texas, and New York, have large populations. According to the histogram, California is the lone outlier.

1.71

```
0 | 44
0 | 555678888999
1 | 012223
1 | 68
2 | 1
2 | 5
3 | 244
3 |
4 |
4 | 9
5 | 002233
5 |
6 | 0113
6 | 5578
7 | 0002
7 | 678
8 | 00
```

The distribution has two distinct peaks, in the second 0 stem and the first 5 stem. The midpoint is located between the 2 leaf and the first 4 leaf in the first 3 stem. The center does not provide us with information about the distribution's multiple peaks, or, more generally, about the fact that the distribution breaks into several distinct "clusters" of observations.

1.72

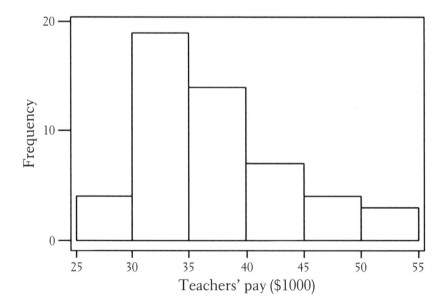

The distribution is skewed to the right. There are no apparent outliers.

1.73 (a)

20.8	Northeast
21.2	Northeast
20.0	Northeast
17.8	Northeast
28.0	Northeast
19.2	Northeast
23.3	Northeast
25.2	Northeast
25.3	Northeast
33.1	South
35.4	South
35.7	South
32.9	South
22.5	South
25.6	South
29.1	South
21.6	South
30.0	South
31.7	South
24.8	South
34.0	South

(b)

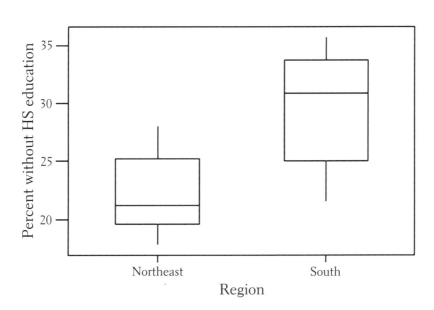

Northeast: Five-number summary: 17.8, 19.6, 21.2, 25.25, 28 (IQR = 5.65)
             Mean and standard deviation: $\bar{x} = 22.31$, $s = 3.34$

Southern: Five-number summary: 21.6, 25.2, 30.85, 33.55, 35.7 (IQR = 8.35)
             Mean and standard deviation: $\bar{x} = 29.7$, $s = 4.97$

The percent of individuals without high school diplomas is distinctly higher in the Southern states than it is in the Northeastern states. The first quartile of the Southern states and the third quartile of the Northeastern states are virtually the same. The shapes of the distribution are somewhat different, with the boxplots indicating that the Southern distribution is skewed left and the Northeastern distribution is skewed right.

The Normal Distributions

2.1 There are many correct drawings. Here are two possibilities:

(a)
(b)

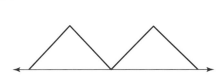 

2.2 (a) The area under the curve is a rectangle with height 1 and width 1. Thus the total area under the curve = $1 \times 1 = 1$.

(b) 20%. (The region is a rectangle with height 1 and base width 0.2; hence the area is 0.2.)

(c) 60%.

(d) 50%.

(e) Mean = $^1/_2$ or 0.5, the "balance point" of the density curve.

2.3 (a) Total area under curve = area of triangle + area of 2 rectangles = $^1/_2(1)(.4) + (.4)(1) + (.4)(1)$ $= 0.2 + 0.4 + 0.4 = 1$.

(b) 0.2.

(c) 0.6.

(d) 0.35.

(e) The median is the "equal-areas" point. By (d), the area between 0 and 0.2 is 0.35. The area between 0.4 and 0.8 is 0.4. Thus the "equal-areas" point must lie between 0.2 and 0.4.

2.4 (a) Mean C, median B; (b) mean A, median A; (c) mean A, median B.

2.5 The uniform distribution. Each of the 6 bars should have a height of 20.

2.6

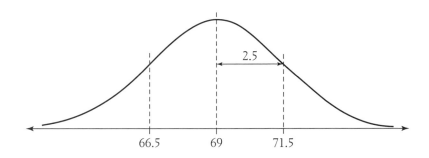

2.7 (a) 2.5% (this is 2 standard deviations above the mean). (b) 69 ± 5; that is, 64 to 74 inches. (c) 16%. (d) 84%. The area to the left of X = 71.5 under the N(69, 2.5) curve is 0.84.

2.8 (a) 50%. (b) 2.5%. (c) 110 ± 50, or 60 to 160.

2.9 (a) 50. The mean (center of the distribution in this case) is 64.5, so 50% of the area to the left of 64.5 under the N(64.5, 2.5) curve. Parts (b) through (d) use similar reasoning. (b) 2.5 (c) 84 (d) 99.85.

2.10 Answers vary.

2.11 Approximately 0.2 (for the tall one) and 0.5.

2.12 (a) 16%. (b) 84th percentile. The area to the left of X = 23.9 under the N(22.8, 1.1) curve is 0.84. (c) 68%.

2.13 (a) 266 ± 32, or 234 to 298 days. (b) Less than 234 days. (c) More than 298 days (2 standard deviations to the right of the mean).

2.14 (a) By the 68-95-99.7 rule, approximately 16% of the scores lie below $\mu - 1\sigma = 110 - 25 = 85$. The TI-83 command shadeNorm (−1000, 85, 110, 25) reports a lower left tail area of .158655. (b) The 84th percentile is the area under the N(110,25) curve to the left of $\mu + 1\sigma = 110 + 25 = 135$. The command shadeNorm (−1000, 135, 110, 25) reports an area of .841345. The 97.5 percentile is the area to the left of $\mu + 2\sigma = 110 + 50 = 160$. ShadeNorm (−1000, 160, 110, 25) reports an area of .97725.

2.15 (a)

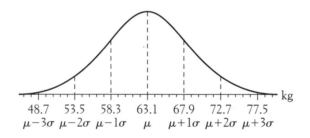

(b) 68%: (58.3, 67.9). 95%: (53.5, 72.7). 99.7%: (48.7, 77.5).

2.16 (a)

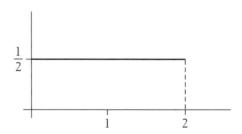

(b) 50% of the outcomes are less than 1. (c) Because the distribution is symmetric, the median is 1, $Q_1 = .5$, and $Q_3 = 1.5$. (d) For $0.5 < X < 1.3$, the proportion of outcomes is 0.8 (1/2) = 0.4.

2.17 (a) Outcomes around 25 are more likely. (d) The distribution should be roughly symmetric with center at about 25, single peaked at the center, standard deviation about 3.5, and few or no outliers. The normal density curve should fit this histogram well.

2.18 The appearance of the graph indicates that the functions Y_1 and Y_2 are identical; they both represent the density curve of the standard normal probability distribution.

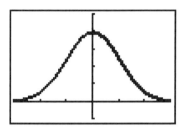

2.19 Eleanor's z-score is $(680 - 500)/100 = 1.8$; Gerald's is $(27 - 18)/6 = 1.5$. Eleanor's score is higher.

2.20 Cobb: $z = (.420 - .266) \div .0371 = 4.15$; Williams: $z = 4.26$; Brett: $z = 4.07$. Williams' z-score is highest.

2.21

(a) 0.9978.

(b) $1 - 0.9978 = 0.0022$.

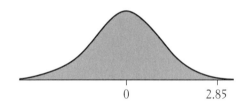

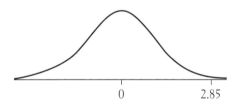

(c) $1 - 0.0485 = 0.9515$.

(d) $0.9978 - 0.0485 = 0.9493$.

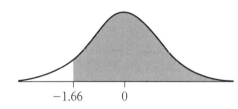

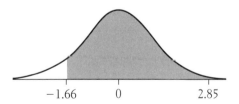

2.22

(a) About -0.675 (-0.67449).

(b) About 0.25 (0.253347).

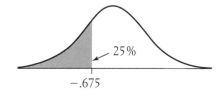

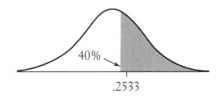

2.23 (a) $z = 3/2.5 = 1.20$; $1 - 0.8849 = 0.1151$; about 11.5%. (b) About 88.5%. (c) 72.2 inches.

2.24 (a) 65.5%. (b) 5.5%. (c) About 127 (or more).

2.25 (a) Area under the density curve of $X = N(0.37, 0.04)$ to the right of 0.40 = .2266.
(b) Area under the density curve of $X = N(0.37, 0.04)$ between 0.40 and 0.50 = .2261.
(c) Proportion in (a) changes to .6915; proportion in (b) changes to .6915.

2.26 (a)

```
48 | 8
49 |
50 | 7
51 | 0
52 | 6799
53 | 04469
54 | 2467
55 | 03578
56 | 12358
57 | 59
58 | 5
```

(b) $\bar{x} = 5.4479$ and $s = 0.22095$. About 75.8% (22 out of 29) lie within one standard deviation of $\bar{x}$, while 96.6% (28/29) lie within two standard deviations.

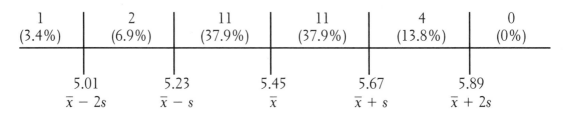

(c)

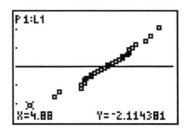

The linearity of the normal probability plot indicates an approximately normal distribution.

2.27 (a)

```
 9 | 4
10 |
11 |
12 | 12346
13 | 22225668
14 | 3679
15 | 237788
16 | 122446788
17 | 688
18 | 23677
19 | 17
20 |
21 |
22 | 8
```

The distribution is approximately symmetric with a peak at the 16 stem. Outliers exist at 9.4 and 22.8.

(b) $\bar{x}$ = 15.586, Median = 15.75. The similarity of these measures reflects the symmetry observed in part (a).

(c) $\bar{x}$ = 15.586, s = 2.55. About 68.2% (30 out of 44) fall within one standard deviation of $\bar{x}$, and about 95.45% (42 out of 44) fall within two standard deviations of $\bar{x}$. All data values (100%, or 44 out of 44) fall within three standard deviations of $\bar{x}$.

1	5	14	16	7	1
(2.3%)	(11.4%)	(31.8%)	(36.4%)	(15.9%)	(2.3%)

10.486	13.036	15.586	18.136	20.686
$\bar{x} - 2s$	$\bar{x} - s$	$\bar{x}$	$\bar{x} + s$	$\bar{x} + 2s$

(d) Window: $[5, 25]_5$ by $[-3, 3]_1$. The plot is strongly linear, with the smallest and largest length observations, 9.4 and 22.8, being the most significant deviations from linearity.

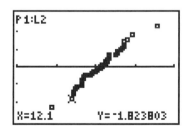

(e) The data appear to be approximately normal. The data satisfy the 68-95-99.7 rule and the shape of the distribution is similar to that of a normal distribution. Also, the strong linearity of the normal probability plot suggests that the data were drawn from a normal distribution.

2.28 (a) 0.0122. (b) 0.9878. (c) 0.0384. (d) 0.9494.

2.29 (a) About 0.52. (b) About -1.04. (c) About 0.84. (d) About -1.28.

2.30 (a) $12\% \pm 2(16.5\%) = -21\%$ to 45%. (b) About 0.23: $x < 0\%$ corresponds to $z < -0.73$, for which Table A gives 0.2335 (software gives 0.2327). (c) About 0.215: $x \geq 25\%$ corresponds to $z \geq 0.79$, for which Table A gives 0.2148 (software gives 0.2154).

2.31 (a) About 5.21%. (b) About 55%. (c) Approximately 279 days or longer.

2.32 (a) A score of $x = 130$ is $z = 2$ standard deviations above the mean, so about 2.5% of 1932 children had very superior scores (2.28% if using Table A). (b) $x > 130$ corresponds to $z > 0.67$, for which Table A gives 0.2514—about 25%.

2.33 (a) At about ± 0.675. (b) For any normal distribution, the quartiles are ± 0.675 standard deviations from the mean; for human pregnancies, the quartiles are 266 ± 10.8, or 255.2 and 276.8.

2.34 (a) At about ± 1.28. (b) 64.5 ± 3.2, or 61.3 to 67.7.

2.35 (a) $\bar{x} = 1.442$, $s = 0.3035$, Median $= 1.45$. The mean and median are virtually identical; the distribution is roughly symmetric.

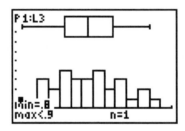

(b) Approximately 67% (20 out of 30) of all observations lie within one standard deviation of the mean. 100% of the observations lie within two and within three standard deviations.

0 (0%)		5 (16.7%)		10 (33.3%)		10 (33.3%)		5 (16.7%)		0 (0%)
	0.835 $\bar{x} - 2s$		1.1385 $\bar{x} - s$		1.442 $\bar{x}$		1.7455 $\bar{x} + s$		2.049 $\bar{x} + 2s$	

(c) The normal probability plot is strongly linear, indicating that the data fit a normal distribution well.

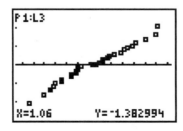

(d) The data appear to be approximately normal. The shape of the distribution is symmetric, the data approximately obey the 68-95-99.7 rule, and the normal probability plot suggests that a normal model is appropriate.

2.36 The normal probability plot shows a strong linear trend. The presidents' ages are approximately normally distributed.

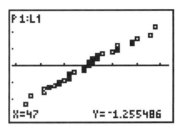

2.37 The mean and standard deviation of a data set of standardized values should be 0 and 1 respectively.

 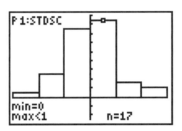

The command 1-Var Stats LSTDSC produces the following screen:

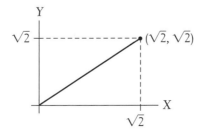

and confirms that the mean is 0 and the standard deviation is 1.

2.38 (a)

(b) If X is the coordinate for which the area between 0 and X under the curve is .5, then 1/2 X base X height = $X^2/2$ = 1/2. Solving for X: $X = 1$. The median is 1. The same approach shows that $Q_1 = \sqrt{.5} = .707$ and $Q_3 = \sqrt{1.5} = 1.225$. (c) The mean will lie to the left of the median (1) because the density curve is skewed left. (d) Area = (.5)(.5)/2 = .125, so 12.5% of the observations lie below 0.5. None (0%) of the observations lie above 1.5.

2.39 (a)

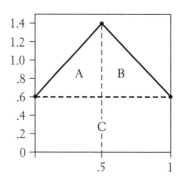

(b) $M = 0.5$, $Q_1 = .3$, $Q_3 = .7$. (c) 25.2%. (d) 49.6%.

2.40 Joey's scoring "in the 97th percentile" on the reading test means that Joey scored as well as or better than 97% of all students who took the reading test and scored worse than only 3%. His scoring in the 72nd percentile on the math portion of the test means that he scored as well as or better than 72% of all students who took the math test and worse than 28%. That is, Joey did better on the reading test, relative to his peers, than he did on the math test.

2.41 (a) 0.8997.   (b) 0.6628.

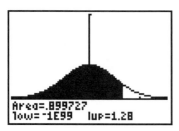   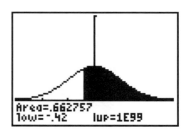

(c) 0.5625.   (d) 0.6628.

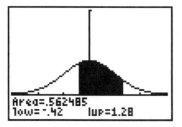   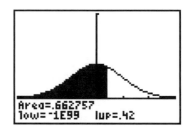

2.42 (a) 98th percentile of $Z = 2.054$. (b) 78th percentile of $Z = 0.7722$.

2.43 $Q_1 = $ 25th percentile of $N(170, 30) = 149.765$; $Q_3 = $ 75th percentile of $N(170, 30) = 190.235$. Median = 50th percentile of $N(170, 30) = 170$.

2.44 (a) $X < 20$ corresponds to $Z < (20 - 25)/5 = -1.00$. The relative frequency is .1587. Alternatively, normalcdf $(-1E99, 20, 25, 5) = .1587$. (b) $X < 10$ corresponds to $Z < (10 - 25)/5 = -3.00$. The relative frequency is .0013, normalcdf $(-1E99, 10, 25, 5) = .00135$. (c) The top quarter corresponds to $z = .675$. Solving $.675 = (x - 25)/5$ gives 28.38.

2.45 13.

2.46 The proportion scoring below 1.7 is about 0.052; the proportion between 1.7 and 2.1 is about 0.078.

2.47 Soldiers whose head circumference is outside the range 22.8 ± 1.81, approximately, less than 21 inches or greater than 24.6 inches.

2.48 Those scoring at least 3.42 are in the "most Anglo/English" 30%; those scoring less than 2.58 make up the "most Mexican/Spanish" 30%.

2.49 (a) Using the window dimensions shown, the histogram shows a distribution that is fairly symmetric with no obvious outliers (a boxplot shows that 10.17 is an outlier). The mean (8.40) is approximately equal to the median (8.42), an indication of symmetry.

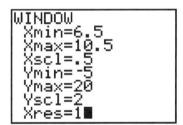
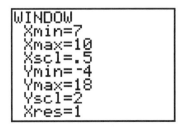
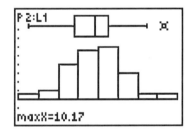

(b) We remove 6.75 (the day after Thanksgiving), and the two largest times, 9.75 and 10.17 (icy roads and a delay due to a traffic accident). The remaining data are slightly skewed left; the mean (8.36) is less than the median (8.42).

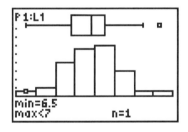
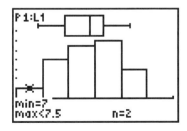
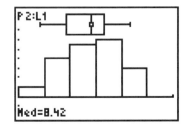

2.50 The normal probability plot for the weight gain for the chicks in the control group (normal corn) is:

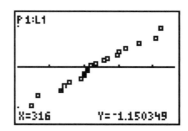

where the observed weight gain is on the *x*-axis and the standardized value is on the *y*-axis. The normal probability plot for the experimental group of chicks fed the lysine added diet is:

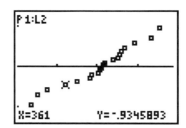

Both plots show a fairly linear pattern of points, so we conclude that both distributions are approximately normal. Thus it is reasonable to use *x* and *s* for the center and spread of the distributions.

2.51 (a) 0.2525. 25.25% of all children have scores greater than 110.

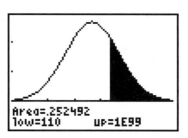

(b) 0.1587. 15.87% of all children have scores lower than 85.

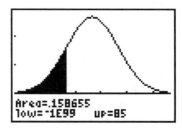

(c) Using the original distribution $N(100, 15)$: Calculate and display the area under the density curve of $N(100, 15)$ between 70 and 130. We obtain an area of 0.9545.

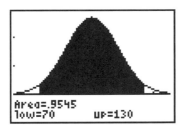

Using the 68-95-99.7 rule: By this rule, 95% of all observations of N(100, 15) must fall within two standard deviations of the mean. The related area is 0.9500.

2.52 **Exercise 2.28:**   `normalcdf (-1E99, -2.25, 0, 1) = 0.0122`
                            `normalcdf (-2.25, 1E99, 0, 1) = 0.9878`
                            `normalcdf (1.77, 1E99, 0, 1) = 0.0384`
                            `normalcdf (-2.25, 1.77, 0, 1) = 0.9494`

   Exercise 2.41:   `normalcdf (-1E99, 1.28, 0, 1) = 0.8997`
                            `normalcdf (-0.42, 1E99, 0, 1) = 0.6628`
                            `normalcdf (-0.42, 1.28, 0, 1) = 0.5623`
                            `normalcdf (-1E99, 0.42, 0, 1) = 0.6628`

2.53 No. Within 4 standard deviations is .999937 area. Going out to 5 standard deviations gives area .999999, which rounds to 1 for 4 decimal place accuracy.

2.54 **Exercise 2.29:**   `invNorm (0.70, 0, 1) = 0.5244 or 0.52`
                            `invNorm (0.15, 0, 1) = -1.0364 or -1.04`
                            `invNorm (0.80, 0, 1) = 0.8416 or 0.84`
                            `invNorm (0.10, 0, 1) = -1.282 or -1.28`

   Exercise 2.42:   `invNorm (0.98, 0, 1) = 2.054 or 2.05`
                            `invNorm (0.78, 0, 1) = 0.7722 or 0.77`

Examining Relationships

3

3.1 (a) Time spent studying is explanatory; the grade is the response variable. (b) Explore the relationship; there is no reason to view one or the other as explanatory. (c) Rainfall is explanatory; crop yield is the response variable. (d) Explore the relationship. (e) The father's class is explanatory; the son's class is the response variable.

3.2 Height at age six is explanatory, and height at age 16 is the response variable. Both are quantitative.

3.3 Sex is explanatory, and political preference in the last election is the response. Both are categorical.

3.4 "Treatment"—old or new—is the (categorical) explanatory variable. Survival time is the (quantitative) response variable.

3.5 The variables are: SAT math score; SAT verbal score. There is no explanatory/response relationship. Both variables are quantitative.

3.6 (a) Explanatory variable = number of powerboat registrations.

(b)

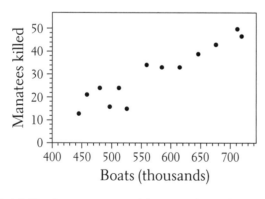

The plot shows a moderately strong linear relationship. As registrations increase, the number of manatee deaths also tends to increase.

3.7 (a) Explanatory variable: number of jet skis in use.

(b)

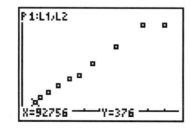

The horizontal axis is "Jet skis in use," and the vertical axis is "Accidents." There is a strong explanatory-response relationship between the number of jet skis in use (explanatory) and the number of accidents (response).

3.8 Answers will vary.

3.9 (a) The variables are positively associated.

(b) The association is moderately linear.

(c) The association is relatively strong. The number of manatees killed can be predicted accurately from the number of powerboat registrations. If the number of registrations remains constant at 719,000, we would expect between 45 and 50 manatees to be killed per year.

3.10 (a) The variables are positively associated; that is, as the number of jet skis in use increases, the number of accidents also increases.

(b) The association is linear.

3.11 (a) Speed is the explanatory variable.

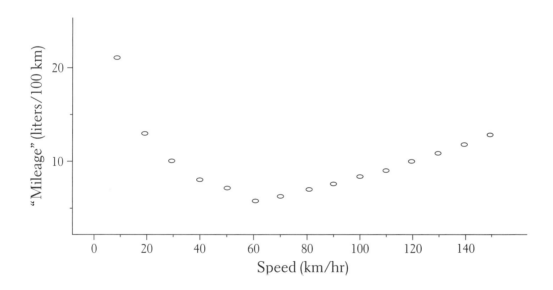

(b) The relationship is curved—low in the middle, higher at the extremes. Since low "mileage" is actually *good* (it means that we use less fuel to travel 100 km), this makes sense: moderate speeds yield the best performance. Note that 60 km/hr is about 37 mph.

(c) Above-average values of "mileage" are found with both low and high values of "speed."

(d) The relationship is very strong—there is little scatter around the curve, and it is very useful for prediction.

3.12 (a) See plot on next page. Body mass is the explanatory variable.

(b) Positive association, linear, moderately strong.

(c) The male subjects' plot can be described in much the same way, though the scatter appears to be greater. The males typically have larger values for both variables.

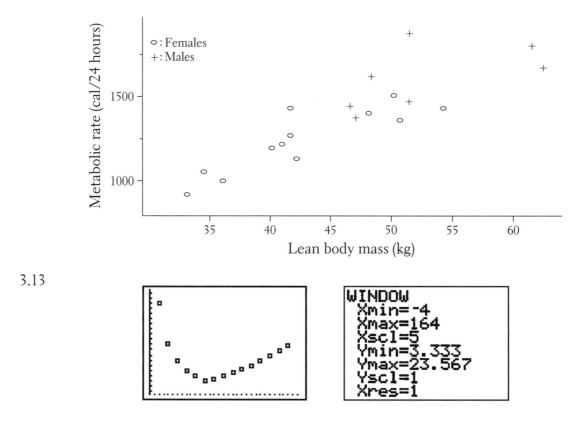

3.13

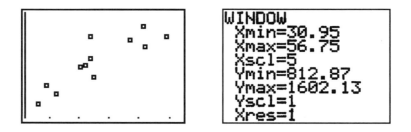

The scatterplot and associated window are shown. The curved association between speed and "mileage" is clearly visible.

3.14 (a) Scatterplot with females only:

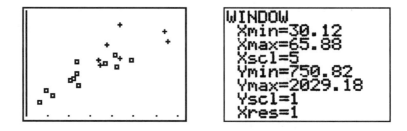

(c) Scatterplot with females (□) and males (+):

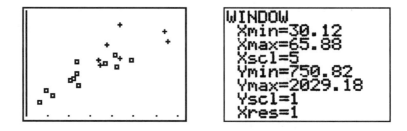

3.15 (a) A positive association between IQ and GPA would mean that students with higher IQs tend to have higher GPAs, and those with lower IQs generally have lower GPAs. The plot does show a positive association.

(b) The relationship is positive, roughly linear, and moderately strong (except for three outliers).

(c) The lowest point on the plot is for a student with an IQ of about 103 and a GPA of about 0.5.

3.16 (a) Lowest: about 107 calories (with about 145 mg of sodium); highest: about 195 calories, with about 510 mg of sodium.

(b) There is a positive association; high-calorie hot dogs tend to be high in salt, and low-calorie hot dogs tend to have low sodium.

(c) The lower left point is an outlier. Ignoring this point, the remaining points seem to fall roughly on a line. The relationship is moderately strong.

3.17 (a) New York's median household income is about $32,800, and the mean per capita income is about $27,500.

(b) The association should be positive since the more money households have, the more money we expect individuals to have. Since the money in a household must be divided among those in the household, we expect household income to be higher than personal income.

(c) Income distributions tend to be right-skewed, which would raise the mean per capita income above the median. In the District of Columbia, this skewness (perhaps combined with small household sizes) overcomes the effect we described in (b).

(d) Alaska's median household income is about $47,900.

(e) Ignoring the outliers, the relationship is strong, positive, and moderately linear.

3.18 (a) Below. Time is explanatory. (b) The association is negative: When time is low, pulse rate is high, and vice versa. This makes sense because finishing faster requires greater effort, and so would raise the pulse rate higher. (c) This is a moderately linear relationship.

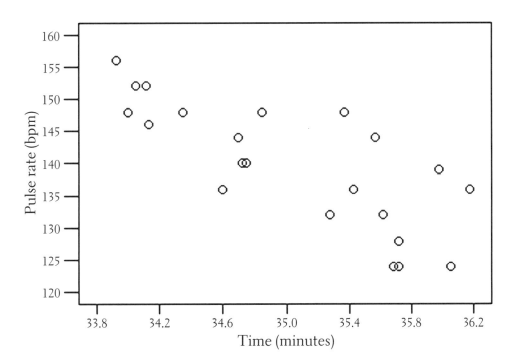

3.19 Since there is no obvious choice for response variable, either could go on the vertical axis. The plot shows a strong positive linear relationship, with no outliers. There appears to be only one species represented.

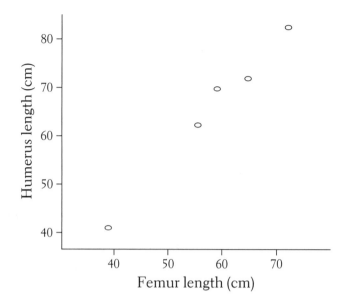

3.20 (a)

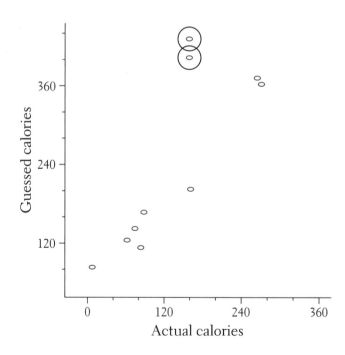

(b) Positive association; approximately linear except for two outliers (circled); spaghetti and snack cake.

3.21 (a) Planting rate is explanatory. (b) See (d). (c) As we might expect from the discussion, the pattern is curved—high in the middle, and lower on the ends. Not linear, and there is neither positive nor negative association. (d) 20,000 plants per acre seems to give the highest average yield.

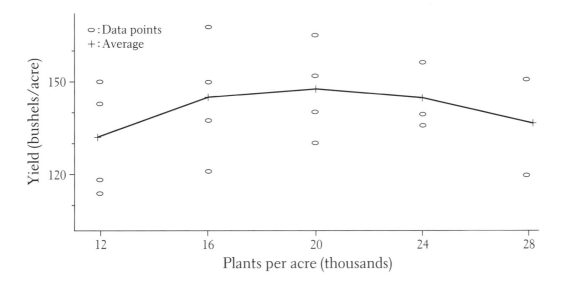

3.22 (a) Below. (b) "The association is negative" means that in general, when one variable takes on a high value, the other is low, and vice versa. That is, in states where many residents did not finish high school, teacher salaries tend to be low, while salaries tend to be higher in states where fewer residents did not finish high school. "The association is weak" means that there are many exceptions to this generalization. (In fact, without the one state in the upper left and the nine states in the lower right, there is little or no association.) (c) Answer will vary from state to state. (d) The outlier is Alaska (13.4% did not finish high school, and the average salary is $49,600). (e) These are the nine states where at least 30% of residents did not finish high school: North Carolina, South Carolina, Louisiana, Tennessee, Alabama, Arkansas, West Virginia, Kentucky, and Mississippi. All are in the southeast quarter of the United States.

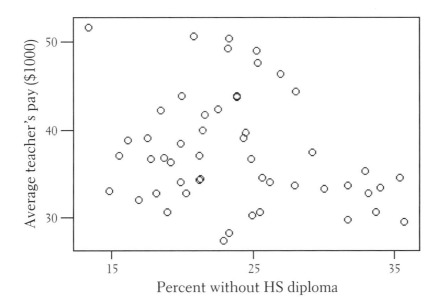

3.23 (a) Plot is on next page. The means are (in the order given) 47.167, 15.667, 31.5, and 14.833. (b) Yellow seems to be the most attractive, and green is second. White and blue are poor attractors.

(c) Positive or negative association make no sense here because color is a categorical variable (what is an "above-average" color?).

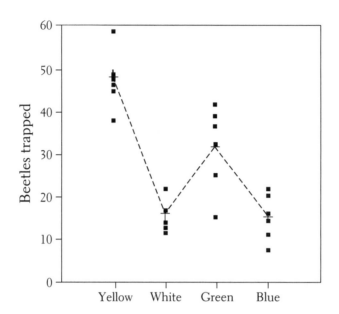

3.24 (a) With x as femur length and y as humerus length: $\bar{x} = 58.2$, $s_x = 13.20$; $\bar{y} = 66.0$, $s_y = 15.89$; $r = 0.994$.

(b) Obviously, the correlation should be identical to the answer obtained in (a).

3.25 (a) The correlation in Figure 3.5 is positive but not near 1; the plot clearly shows a positive association, but with quite a bit of scatter.

(b) The correlation in Figure 3.6 is closer to 1 since the spread is considerably less in this scatterplot.

(c) The outliers in Figure 3.5 weaken the relationship, so dropping them would increase r. The one outlier in Figure 3.6 strengthens that relationship (since the relative scatter about the diagonal line is less when it is present), so the correlation would drop with it removed.

3.26 $r = 1$ (this is a perfect straight line, with a positive slope).

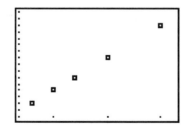

```
WINDOW
 Xmin=26.8
 Xmax=41.2
 Xscl=5
 Ymin=27.96
 Ymax=44.04
 Yscl=1
 Xres=1
```

3.27 (a) See Exercise 3.19 for plot. The plot shows a strong positive linear relationship, with little scatter, so we expect that r is close to 1.

(b) r would not change—it is computed from standardized values, which have no units.

3.28 With x for speed and y for mileage: $\bar{x} = 40$, $s_x = 15.8$; $\bar{y} = 26.8$, $s_y = 2.68$; $r = 0$. Correlation only measures *linear* relationships; this plot shows a strong *non-linear* relationship.

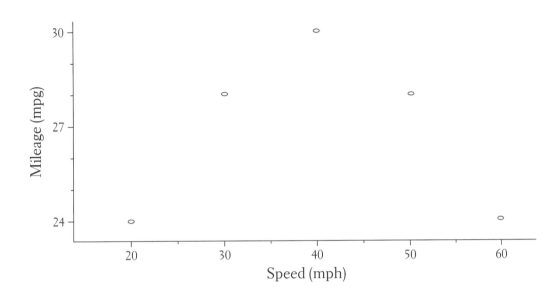

3.29 (a) The correlation is $r \doteq -0.746$, which is consistent with the moderate negative association visible in the plot (see the solution to Exercise 3.18).

(b) Changing the units of measurement does not affect standard scores, and so does not change r.

3.30 (a) See Exercise 3.12 for plot. Both correlations should be positive, but since the men's data seem to be more spread out, it may be slightly smaller. (b) Women: $r_w = 0.87645$; Men: $r_m = 0.59207$. (c) Women: $\bar{x}_w = 43.03$; Men: $\bar{x}_m = 53.10$. The difference in means has no effect on the correlation. (d) There would be no change, since standardized measurements are dimensionless.

3.31 (a) See Exercise 3.20 for plot. $r = 0.82450$. This agrees with the positive association observed in the plot; it is not too close to 1 because of the outliers.

(b) It has no effect on the correlation. If every guess had been 100 calories higher—or 1000, or 1 million—the correlation would have been exactly the same, since the standardized values would be unchanged.

(c) The revised correlation is $r = 0.98374$. The correlation got closer to 1 because without the outliers, the relationship is much stronger.

3.32 (a) On the next page; men are the plus signs, and women are the open circles. $r_{all} \doteq 0.3576$, $r_{men} \doteq 0.4984$, and $r_{women} \doteq 0.3257$.

(b) The points for men are generally located on the right side of the plot, while the women's points are generally on the left. $\bar{x}_{men} \doteq 954,855$ and $\bar{x}_{women} \doteq 862,655$ pixels.

(c) The correlations for men and women suggest that there is a moderate positive association for men and a weak one for women. However, one significant feature of the data that can be observed in the scatterplot is that the sample group was highly stratified; that is, there were 10 men and 10 women with high IQs (at least 130), while the other 10 of each gender had IQs of no more than 103. The men's higher correlation can be attributed partly to the two subjects with large brains and 103 IQs (which are high relative to the low-IQ group). The men's correlation might not remain so high with a larger sample size.

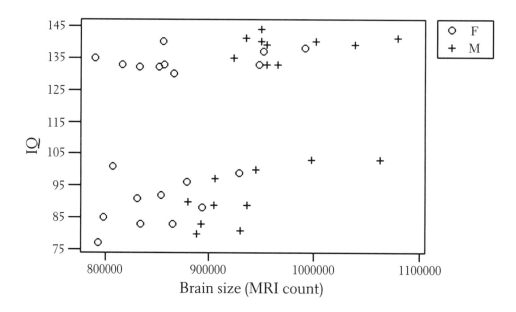

3.33 (a) Here is the scatterplot of the original x-y data (marked with □).

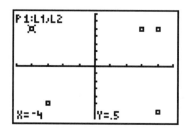

(b) The correlation of the original x-y data is 0.253.

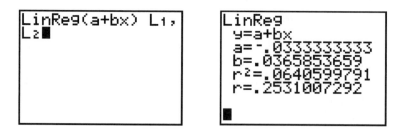

(c) Plot of Y vs. X (□) and Y* vs. X* (+) on same axes:

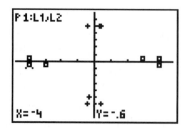

(d) $r = 0.253$. The correlation is the same because changing the units of x, y does not change the value of r.

3.34 The person who wrote the article interpreted a correlation close to 0 as if it were a correlation close to −1. Prof. McDaniel's findings mean there is little linear association between research and teaching—for example, knowing a professor is a good researcher gives little information about whether she is a good or bad teacher.

3.35 (a) Rachel should choose small-cap stocks. Small-cap stocks have a lower correlation with municipal bonds, so the relationship is weaker.

(b) She should look for a negative correlation (although this would also mean that this investment tends to *decrease* when bond prices rise).

3.36 See the solution to Exercise 3.11 for the scatterplot. $r = -0.172$; it is close to zero because the relationship is a curve rather than a line.

3.37 (a) Sex is a categorical variable. (b) r must be between −1 and 1. (c) r should have no units (i.e., it can't be 0.23 *bushel*).

3.38 $\bar{x} = 22.31$, $s_x = 17.74$; $\bar{y} = 5.306$, $s_y = 3.368$; $r = 0.99526$. Except for roundoff error, we again find $b = 0.1890$ and $a = 1.0892$.

3.39 Answers will vary.

3.40 (a) A negative association—the pH decreased (i.e., the acidity increased) over the 150 weeks. (b) The initial pH was 5.4247; the final pH was 4.6350. (c) The slope is −0.0053; the pH decreased by 0.0053 units per week (on the average).

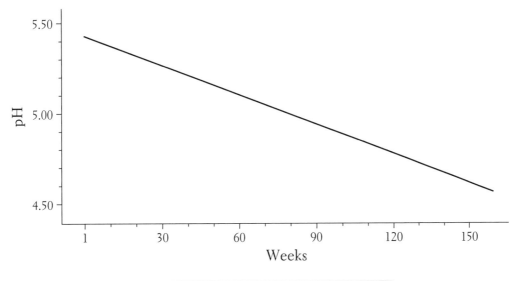

3.41 (a)

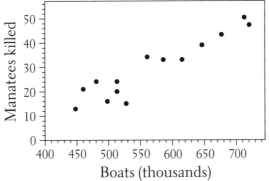

(b) Equation of line: $\hat{y} = -41.43 + .125x$.

(c) When $x = 716$, $y = 48$ dead manatees are predicted.

(d) The additional points are shown as open circles. Two of the points (those for 1992 and 1993) lie below the overall pattern (i.e., there were fewer actual manatee deaths than we might expect), but otherwise there is no strong indication that the measures succeeded.

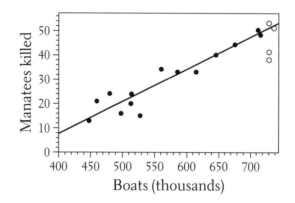

(e) The mean for those years was 42—less than our predicted mean of 48 (which *might* suggest that the measures taken showed some results).

3.42 $r = \sqrt{0.16} = 0.40$ (high attendance goes with high grades, so the correlation must be positive).

3.43 (b) When $x = 34.30$, we predict $y \doteq 147.4$ beats per minute—about 4.6 bpm lower than the actual value.

(c) Regressing time on pulse rate gives the equation Time $= 43.10 - 0.0574$ Pulse, which leads to a predicted time of 34.38 minutes—only 0.08 minutes (4.8 seconds) too high.

(d) The results of a least-squares regression depend on which variable is viewed as explanatory since the line is chosen based on vertical distances from each data point to the line.

3.44 (a) The straight-line relationship explains $r^2 \doteq 35.5\%$ of the variation in yearly changes. (b) $b = r \cdot s_y/s_x \doteq 1.707$; $a = \bar{y} - b\bar{x} \doteq 6.083\%$. The regression equation is $\hat{y} = 6.083\% + 1.707x$. (c) The predicted change is $\hat{y} = 9.07\%$, as it must be, since the regression line must pass through $(\bar{x}, \bar{y})$.

3.45 (a) Stumps should be on the horizontal axis. The plot shows a positive linear association.

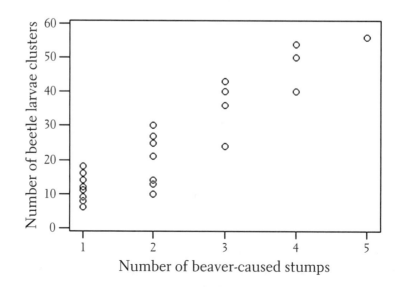

(b) The regression line is $\hat{y} = -1.286 + 11.89x$.

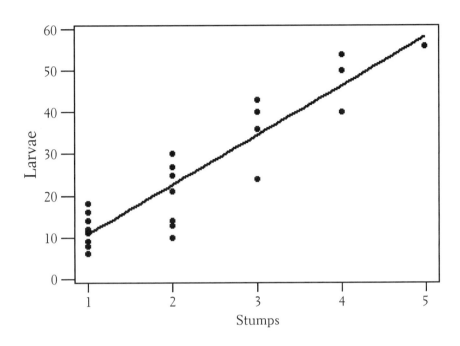

(c) The straight-line relationship explains $r^2 \doteq 83.9\%$ of the variation in beetle larvae.

3.46 (a) Below.

(b) The line is clearly *not* a good predictor of the actual data—it is too high in the middle and too low on each end.

(c) The sum is -0.01—a reasonable discrepancy allowing for roundoff error.

(d) A straight line is not the appropriate model for these data.

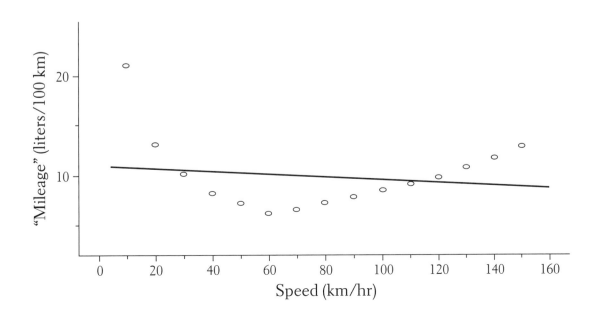

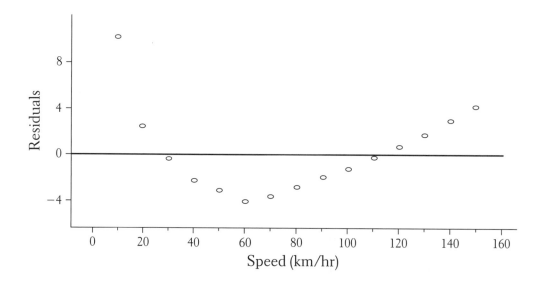

3.47 (a) Below. (b) Let y be "guessed calories" and x be actual calories. Using all points: $\hat{y} = 58.59 + 1.3036x$ (and $r^2 = 0.68$)—the dashed line. Excluding spaghetti and snack cake: $\hat{y}^* = 43.88 + 1.14721x$ (and $r^2 = 0.968$). (c) The two removed points could be called influential, in that when they are included, the regression line passes above every *other* point; after removing them, the new regression line passes through the "middle" of the remaining points.

3.48 (a) Without Child 19, $\hat{y}^* = 109.305 - 1.1933x$. Child 19 might be considered *somewhat* influential, but removing this data point does not change the line substantially.

(b) With all children, $r^2 = 0.410$; without Child 19, $r^2 = 0.572$. With Child 19's high Gesell score removed, there is less scatter around the regression line—more of the variation is explained by the regression.

3.49 (a) See facing page. For this plot $x = $ MASS F, $y = $ MET F.

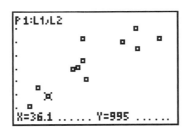

(b) Equation of least-squares line: metabolic rate = 201.1616 + (24.026 × lean body mass). r^2 = .7682, so lean body mass explains about 76.82% of the variation in metabolic rate.

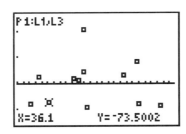

(c) From the residual plot, the line does appear to provide an adequate model. The residuals are scattered about the horizontal axis and no patterns are evident.

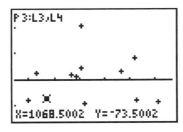

(d) The plots are identical in terms of the relative positions of the data points.

3.50 (a) Graph not shown. (b) $2,500. (c) $y = 500 + 200x$.

3.51 (a) y (weight) = $100 + 40x$ grams. (b) Graph not shown. (c) When $x = 104$, $y = 4260$ grams, or about 9.4 pounds—a rather frightening prospect. The regression line is only reliable for "young" rats; like humans, rats do not grow at a constant rate throughout their entire lives.

3.52 (a) $b = r \cdot s_y/s_x \doteq 0.1010$; $a = \bar{y} - b\bar{x} \doteq -3.557$. The regression equation is $\hat{y} = -3.557 + 0.1010x$. (b) The straight-line relationship explains $r^2 = 40.16\%$ of the variation in GPAs. (c) The predicted GPA is $\hat{y} \doteq 6.85$; the residual is -6.32.

3.53 (a) Since both variables are measured in dollars, the same scale is used on both axes. The plot shows (perhaps) a weak positive association, with one outlier.

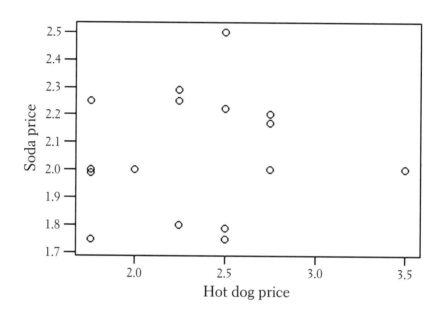

(b) The correlation is $r \doteq 0.1426$; about $r^2 \doteq 2\%$ of soda price variation is explained by a linear relationship with hot dog price.

(c) The regression line is $\hat{y} = 1.9057 + 0.0619x$. The slope is near zero because the relationship is weak; regardless of the hot dog price x, our predicted soda price is near the mean soda price (about \$2.05).

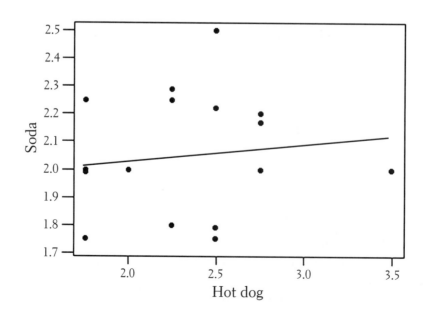

(d) The outlier is the point for the Cardinals, with an extremely expensive hot dog and a soda priced just below average. Without that point, the regression equation is $\hat{y} = 1.7613 + 0.1303x$ (the dashed line). Some might call this point influential since the line is slightly different without it (and r^2 increases to 5%); however, there is not much difference between the two lines over the range of the remaining hot dog prices, so whether we should call this point influential is a matter of opinion.

3.54 (a) Plot below. The correlation is $r \doteq 0.9999$, so recalibration is not necessary. (b) The regression line is $\hat{y} = 1.6571 + 0.1133x$; when $x = 500$ mg/liter, we predict $\hat{y} = 58.31$. This prediction should be very accurate since the relationship is so strong.

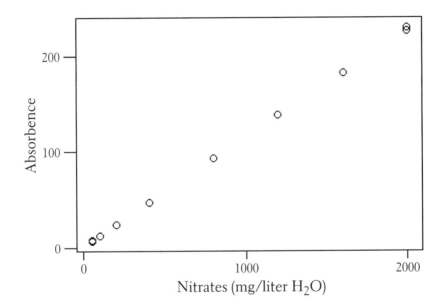

3.55 (a) Below. (b) $\hat{y} = 71.950 + 0.38333x$. (c) When $x = 40$, $\hat{y} = 87.2832$; when $x = 60$, $\hat{y} = 94.9498$. (d) Sarah is growing at about 0.38 cm/month; she should be growing about 0.5 cm each month $\left(0.5 = \frac{6}{60 - 48}\right)$.

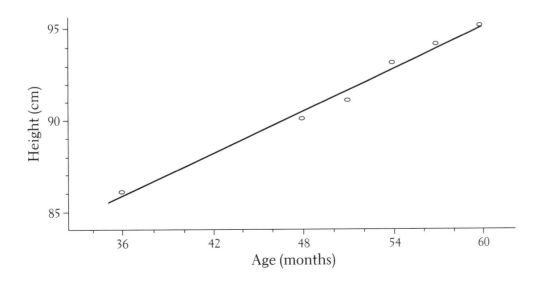

3.56 (a) Below. Since both variables are measured in the same units, the same scale is used on both axes.

(b) $r = 0.463$ and $r^2 = 0.214 = 21.4\%$. There is a positive association between U.S. and overseas returns, but it is not very strong: Knowing the U.S. return accounts for only about 21.4% of the variation in overseas returns.

(c) The regression equation is $\hat{y} = 5.683 + 0.6181x$.

(d) When $x = 33.4\%$, $\hat{y} \doteq 26.3\%$ Since the correlation is so low, the predictions will not be very reliable.

(e) In 1986, the overseas return was 69.4%—over 50 percentage points higher than would be expected. There are no points that look influential.

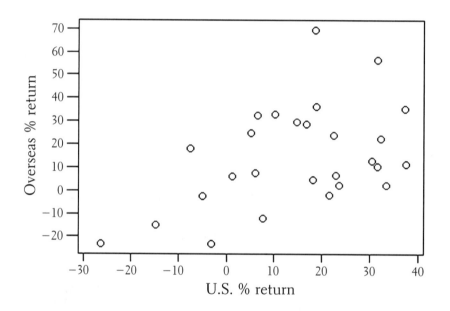

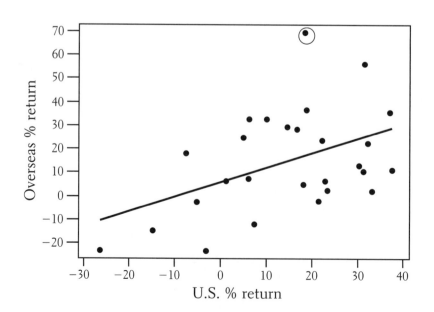

3.57 (a) $b = r \cdot s_y/s_x = (0.6)(8)/(30) = 0.16$; $a = \bar{y} - b\bar{x} = 30.2$. (b) Julie's predicted score is $\hat{y} \doteq 78.2$. (c) $r^2 = 0.36$; only 36% of the variability in y is accounted for by the regression, so the estimate $\hat{y} = 78.2$ could be quite different from the real score.

3.58 When $x = 480$, $\hat{y} = 255.95$ cm, or 100.77 in, or about 8.4 feet!

3.59 (a) Table and plot below.

	Min	Q_1	M	Q_3	Max
U.S.	−26.4%	5.1%	18.2%	30.5%	37.6%
Overseas	−23.4%	2.1%	11.2%	29.6%	69.4%

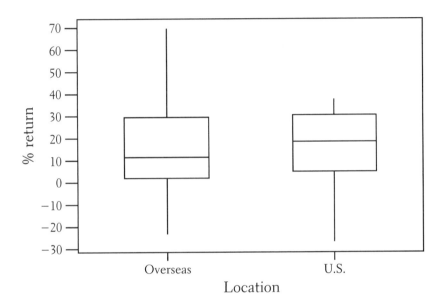

(b) Either answer is defensible: The three middle numbers of the U.S. five-number summary are higher, but the minimum and maximum overseas returns are higher.

(c) Overseas stocks are more volatile—the boxplot is more widely spread. Also, the low U.S. return (−26.4%) appears to be an outlier; not so with the overseas stocks.

3.60 Note that $\bar{y} = 46.6 + 0.41\bar{x}$. We predict that Octavio will score 4.1 points above the mean on the final exam: $\hat{y} = 46.6 + 0.41(\bar{x} + 10) = 46.6 + 0.41\bar{x} + 4.1 = \bar{y} + 4.1$. (Alternatively, since the slope is 0.41, we can observe that an increase of 10 points on the midterm yields an increase of 4.1 on the predicted final exam score.)

3.61 (a)

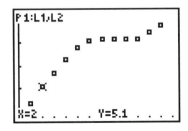

(b) See scatterplot on next page, with line superimposed. Clearly, this line does not fit the data very well; the data show a clearly curved pattern.

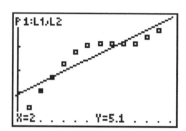

(c) The residuals sum to 0.01 (the result of roundoff error). The residual plot below shows a clearly curved pattern, verifying the result in part (b). The residuals are positive between 3 and 8 months but are negative at all other times.

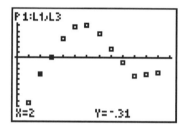

3.62 (a) Two mothers are 57 inches tall; their husbands are 66 and 67 inches tall. (b) The tallest fathers are 74 inches tall; there are three of them, and their wives are 62, 64, and 67 inches tall. (c) There is no clear explanatory variable; either could go on the horizontal axis. (d) The weak positive association indicates that people have *some* tendency to many persons of a similar *relative* height—but it is not an overwhelming tendency. It is weak because there is a great deal of scatter.

3.63 (a) Below. Alcohol from wine should be on the horizontal axis. There is a fairly strong linear relationship. The association is negative: Countries with high wine consumption have fewer heart disease deaths, while low wine consumption tends to go with more deaths from heart disease.

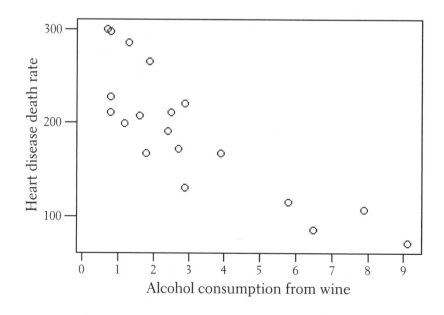

(b)

MINITAB output for least square line:

```
The regression equation is DEATHRT = 261 - 23.0 ALCOHOL
```

The correlation is $r = -0.843$.

(c) The correlation is reasonably close to -1, indicating that predictions of death rate from wine consumption will be fairly accurate. $r^2 = (-0.843)^2 = 0.711$, so about 71.1% of the variation in death rate can be explained by the linear relationship.

(d) Predicted death rate for wine consumption = 4: $261 - (23 \times 4) = 168.8$.

(e) No. Positive r indicates that the least-squares line must have positive slope, negative r indicates that it must have negative slope. The direction of the association and the slope of the least-squares line must always have the same sign.

3.64 (a) The point at the far left of the plot (Alaska) and the point at the extreme right (Florida) are outliers. Alaska may be an outlier because its cold temperatures discourage older residents from remaining in the state. Florida is an outlier because many individuals choose to retire there.

(b) The association is positive and (very) weakly linear.

(c) The correlation without the outliers is $r = 0.267$. Removing the outliers causes the association to be a bit stronger (more linear).

3.65 (a) To three decimal places, the correlations are all approximately 0.816 (r_D actually rounds to 0.817), and the regression lines are all approximately $\hat{y} = 3.000 + 0.500x$. For all four sets, we predict $\hat{y} \doteq 8$ when $x = 10$. (b) Below. (c) For Set A, the use of the regression line seems to be reasonable—the data do seem to have a moderate linear association (albeit with a fair amount of scatter). For Set B, there is an obvious *nonlinear* relationship; we should fit a parabola or other curve. For Set C, the point (13, 12.74) deviates from the (highly linear) pattern of the other points; if we can exclude it, regression would be very useful for prediction. For Set D, the data point with $x = 19$ is a very influential point—the other points alone give no indication of slope for the line. Seeing how widely scattered the y-coordinates of the other points are, we cannot place too much faith in the y-coordinate of the influential point; thus we cannot depend on the slope of the line, and so we cannot depend on the estimate when $x = 10$.

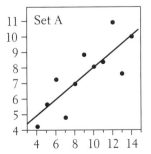 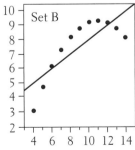 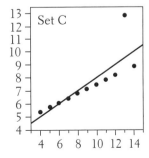 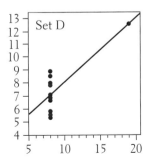

3.66 (a) Shown on next page. Fatal cases are marked with solid circles; those who survived are marked with open circles. (b) There is no clear relationship. (c) Generally, those with short incubation periods are more likely to die. (d) Person 6—the 17-year-old with a short incubation

(20 hours) who survived—merits extra attention. He or she is also the youngest in the group by far. Among the other survivors, one (person 17) had an incubation period of 28 hours, and the rest had incubation periods of 43 hours or more.

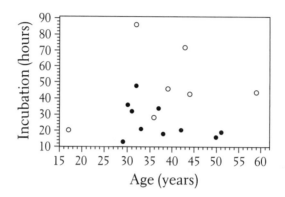

3.67 The plot shows an apparent negative association between nematode count and seedling growth. The correlation supports this: $r = -0.78067$. This also indicates that about 61% of the variation in growth can be accounted for by a linear relationship with nematode count.

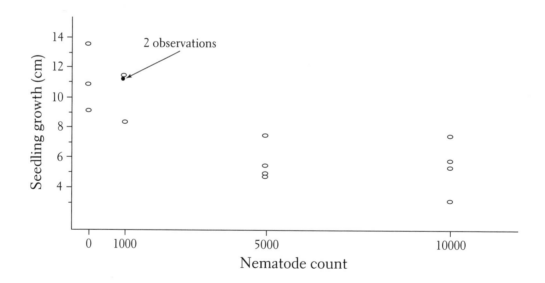

3.68 (a) $b = r \cdot s_y/s_x \doteq 1.1694$; $a = \bar{y} - b\bar{x} \doteq 0.3531$. The regression equation is $\hat{y} = 0.3531 + 1.1694x$; it explains $r^2 \doteq 27.6\%$ of the volatility in Philip Morris stock.

(b) On the average, for every percentage-point rise in the S&P monthly return, Philip Morris stock returns rise about 1.17 percentage points. (And similarly, Philip Morris returns fall 1.17% for each 1% drop in the S&P index return.)

(c) When the market is rising, the investor would like to earn money faster than the prevailing rate, and so prefers beta > 1. When the market falls, returns on stocks with beta < 1 will drop more slowly than the prevailing rate.

3.69 $b = 0.54$ (and $a = 33.67$). For $x = 67$ inches, we estimate $\hat{y} = 69.85$ inches.

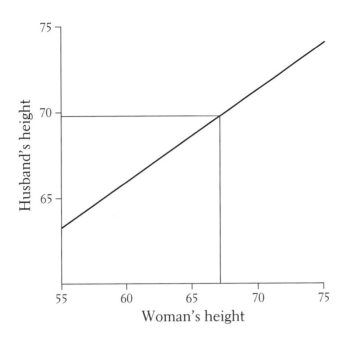

3.70 (a) Below. (b) The slope is close to 1—meaning that the strength after 28 days is *approximately* (strength after one week) plus 1389 psi. In other words, we expect the extra three weeks to add about 1400 psi of strength to the concrete. (c) 4557 psi.

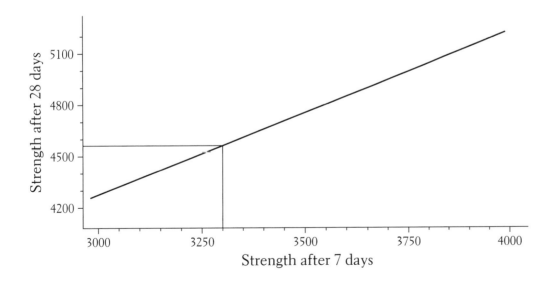

3.71 (a) On next page. (b) There is a very strong positive linear relationship; $r = 0.9990$. (c) Regression line: $\hat{y} = 1.76608 + 0.080284x$ (y is steps/second, x is speed). (d) $r^2 = 0.998$, so nearly all the variation (99.8% of it) in steps taken per second is explained by the linear relationship. (e) The regression line would be different (as in Example 3.11), because the line in (c) is based on minimizing the sum of the squared *vertical* distances on the graph. This new regression would minimize the squared *horizontal* distances (for the graph shown). r^2 would remain the same, however.

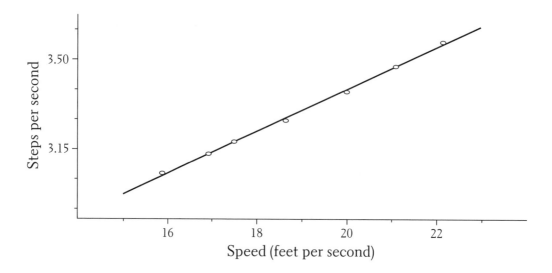

3.72 (a) No. Consider the following example: Start with points (1, 1) and (2, 2). Then add the influential point (0, 4).

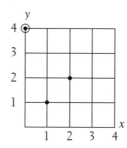

The addition of (0, 4) makes *r negative*; the correlation with the original two points was positive. (b) No. Consider the following example: Start with the set of points (1, 1), (1, 2), (2, 1.1), and (2, 2). Then add the influential point (10, 10).

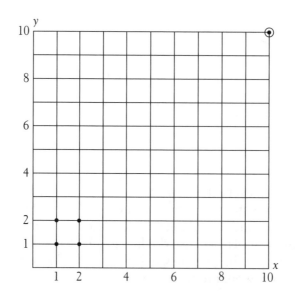

The regression line with the added point is $\hat{y} = .08 + .981x$. The line with the point removed is $\hat{y} = 1.45 + .05x$. Both a and b change dramatically!

3.73 (a) Franklin is marked with a + (in the lower left corner). (b) There is a moderately strong positive linear association. (It turns out that $r^2 = 87.0\%$.) There are no really extreme observations, though Bank 9 did rather well. Franklin does not look out of place. (c) $\hat{y} = 7.573 + 4.9872x$. (d) Franklin's predicted income was $\hat{y} = 26.5$ million dollars—almost twice the actual income. The residual is -12.7.

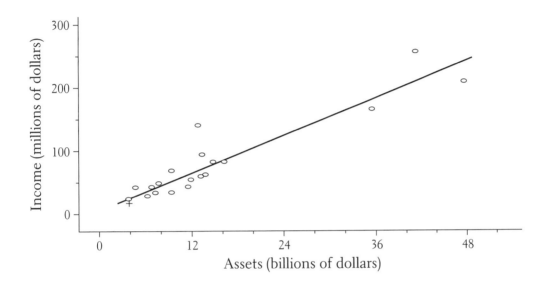

3.74 (a) See Exercise 3.72 (a). (b) See Exercise 3.72 (b).

3.75 (a) The men's times (the plot of L_2 on L_1) have been steadily decreasing for the last 100 years.

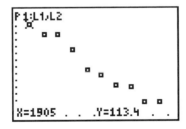

Likewise, the women's times (in the plot of L_4 on L_3) have also been decreasing steadily.

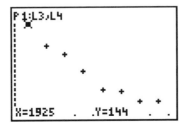

A straight line appears to be a reasonable model for the men's times; the correlation for the men is $r = .983$.

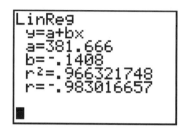
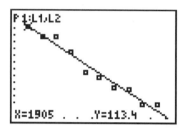

A straight line also appears to be a reasonable model for the women's times; the correlation for the women is $r = -.9687$.

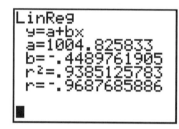
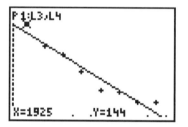

In order to plot both scatterplots on the same axes, we deleted the first two men's times (1905). The resulting plot suggests that while the women's scores were decreasing faster than the men's times from 1925 through 1965, the times for both sexes from 1975 have tended to flatten out. In fact, for the period 1925 to 1995, the best model may not be a straight line, but rather a curve, such as an exponential function. In any event, it is not clear, as Whitt and Ward suggest, that women will soon outrun men.

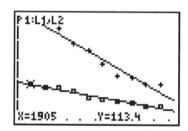

To find out where the two regression lines intersect, set the Window as shown and plot the two lines together. Selecting 2nd / CALC / 5:Intersect yields the point (2002.09, 96.956).

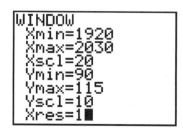
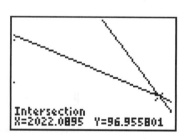

3.76 (a) The circled observation is (716, 35).

(b) A standardized residual of -2.08 is a fairly unusual occurrence. The chances of a residual falling outside the limits -2 to 2 are only 5% by the 68-95-99.7 rule. Thus we could expect to see a residual this extreme only about 5% of the time.

3.77 (a) The scatterplot (previously obtained in Exercise 3.7) and the correlation of $r = 0.932$ suggest that there is a strong linear relationship between the number of jet-ski registrations and the number of jet-ski fatalities. The MINITAB output for the least-squares line is:

```
The regression equation is FATALITIES = 8.00 +0.000066 JETSKIS
```

A MINITAB fitted-line plot yields the following:

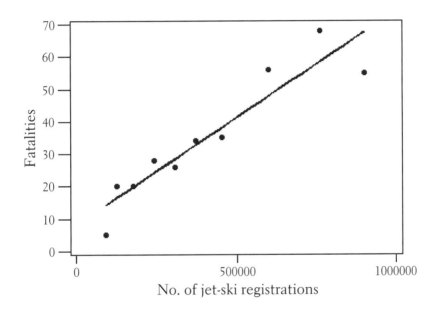

The pattern of residuals is sufficiently random to suggest that the line does a good job of modeling the relationship.

(b) Answers vary.

4

More on Two-Variable Data

4.1 (a) Since 2.54 cm = 1 inch, inches are changed to centimeters by multiplying by 2.54. Letting x = height in inches and y = height in centimeters, the transformation is $y = 2.54x$, which is monotonic increasing. (b) If x = typing speed in words per minute, then $\frac{1}{60}x$ = typing speed in words per second and $y = \frac{60}{x}$ = number of seconds needed to type a word. The transformation is $y = \frac{60}{x}$, which is monotonic decreasing (we can ignore the case $x < 0$ here). (c) Letting x = diameter and y = circumference, the transformation is $y = \pi x$, which is monotonic increasing. (d) Letting x = the time required to play the piece, the transformation is $y = (x - 5)^2$, which is not monotonic.

4.2 Monotonic increasing: intervals 0 to $\frac{\pi}{2}$ and $\frac{3\pi}{2}$ to 2π. Monotonic decreasing: interval $\frac{\pi}{2}$ to $\frac{3\pi}{2}$.

4.3 Weight = c_1 (height)^3 and strength = c_2 (height)^2; therefore, strength = c (weight)$^{2/3}$, where c is a constant.

4.4 A graph of the power law $y = x^{2/3}$ (see below) shows that strength does *not* increase linearly with body weight, as would be the case if a person 1 million times as heavy as an ant *could* lift 1 million times more than the ant. Rather, strength increases more slowly. For example, if weight is multiplied by 1000, strength will increase by a factor of $(1000)^{2/3} = 100$.

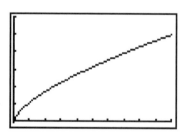

4.5 Let y = average heart rate and x = body weight. Kleiber's law says that total energy consumed is proportional to the three-fourths power of body weight, that is, Energy = $c_1 x^{3/4}$. But total energy consumed is also proportional to the product of the volume of blood pumped by the heart and the heart rate, that is, Energy = c_2 (volume) y. Furthermore, the volume of blood pumped by the heart is proportional to body weight, that is, Volume = $c_3 x$. Putting these three equations together yields $c_1 x^{3/4} = c_2$ (volume) $y = c_2 (c_3 x) y$. Solving for y, we obtain

$$y = \frac{c_1 x^{3/4}}{c_2 c_3 x} = cx^{-1/4}.$$

4.6 (a)

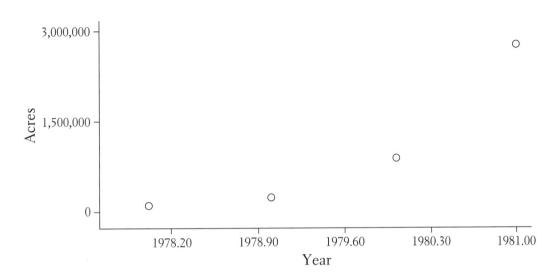

(b) The ratios are 226,260/63,042 = 3.59, 907,075/226,260 = 4.01, and 2,826,095/907,075 = 3.12.

(c) log y yields 4.7996, 5.3546, 5.9576, and 6.4512. Here is the plot of log y vs. x.

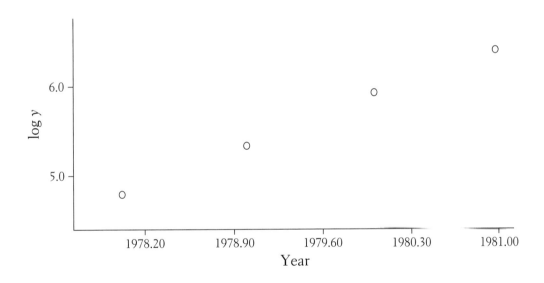

(d) Defining list L_3 on the TI-83 to be log(L_2) and then performing linear regression on (L_1, L_3) where L_1 holds the year (x) and L_3 holds the log acreage yield the least squares line log $\hat{y} = -1094.507 + .55577x$. For comparison, Minitab reports

```
The regression equation is
log y = -1095 + 0.556 Year
Predictor          Coef        Stdev      t-ratio          p
Constant        -1094.51       29.26       -37.41      0.001
Year             0.55577      0.01478       37.60      0.001

s = 0.03305      R-sq = 99.9%      R-sq(adj) = 99.8%
```

(e)

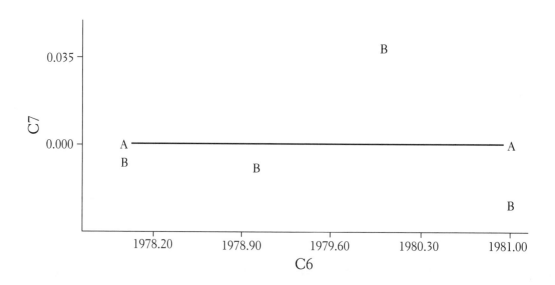

A = C7 vs. C6
B = RESI2 vs. Year

The residual plot of the transformed data shows no clear pattern, so the line is a reasonable model for these points.

(f) $\log \hat{y} = -1094.51 + .5558x$, so $10^{-\log \hat{y}} = 10^{(-1694.51 + .5558x)}$, and so $\hat{y} = 10^{(-1094.51 + .5558x)}$. With the regression equation for $\log y$ on x installed as Y_1 on the TI-83, define $Y_2 = 10^{\wedge}(Y_1)$. Then plot the original scatterplot and Y_2 together.

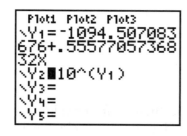 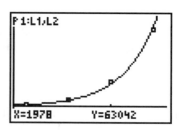

(g) The predicted number of acres defoliated in 1982 is the exponential function evaluated at 1982, which gives 10,719,964.92 acres.

4.7 (a) A scatterplot of the data (see facing page) shows the characteristic exponential growth shape, and a plot of the ratios y_{n+1}/y_n shows some initial variability, but settles down about the line $y = 2.7$.

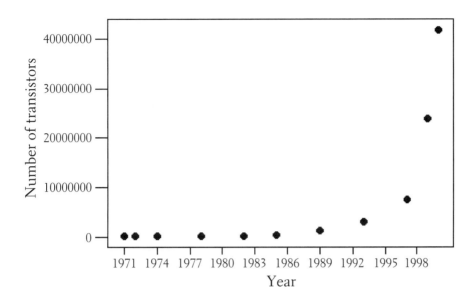

(b) Plotting log y vs. x (see below) shows a straight-line pattern. For this data, $r^2 = .995$.

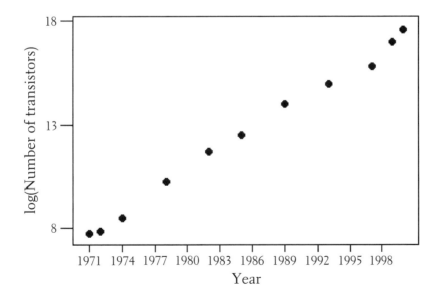

Regressing log y on x and then inspecting the residual plot for the transformed data (see next page) shows no clear pattern, so a straight line is an appropriate model for the transformed data. We conclude that the number of transistors on a chip is growing exponentially.

Residuals Versus DATE
(response is LOGTRANS)

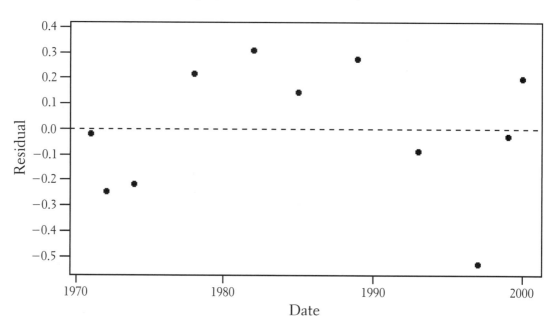

4.8 (a) MINITAB output

```
The regression equation is
LOGTRANS = 3.22 + 0.144 DATE
```

(Here, DATE = number of years after 1970.)

(b) If Moore's law holds exactly, then the number of transistors in year x after 1970 will quadruple every three years. In other words, if $y = ab^x$, then $y(1) = 2250 = ab$ and $y(4) = 9000 = ab^4$. Solving for a, b yields $a \approx 1417.41$, $b \approx 1.5874$, so the Moore's law model is $y = 1417.41\,(1.5874)^x$. The MINITAB output for the regression equation of the transformed data is:

```
The regression equation is
LOGMOORE = 3.15 + 0.201 DATE
```

Since the slope of the regression line is larger for the Moore's law model than it is for the actual data, we can conclude that the actual transistor counts have grown more slowly than Moore's law suggests.

4.9 $2^{1 \times 4} = 16$, $2^{5 \times 4} = 1,048,576$.

4.10 (a) According to the claim, the number of children killed doubled every year after 1950.

Year	1951	1952	1953	1954	1955	1956	1957	1958	1959	1960
# children killed	2	4	8	16	32	64	128	256	512	1024

(b)

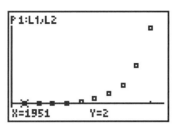

(c) If x = number of years after 1950, then y = the number of children killed x years after $1950 = 2^x$. At $x = 45$, $y = 2^{45} = 3.52 \times 10^{13}$, or 35,200,000,000,000 (35 trillion!).

(d)

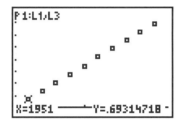

(e) $b \approx 0.3$, $a \approx -587$ from the graph in (d). $\log y$ is therefore related to the year x by the equation $\log y = -587 + 0.3x$. At $x = 1995$, $\log y = -587 + 0.3(1995) = 11.5$. By comparison, $\log (3.52 \times 10^{13}) = 13.5465$. The two values are reasonably close.

4.11 (a)

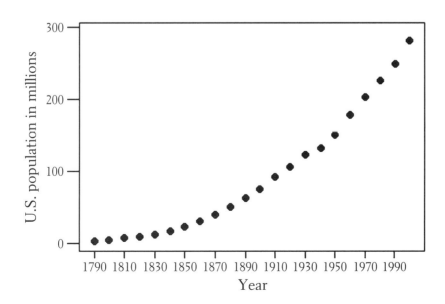

(b) The transformed data (see next page) appear to be linear from 1790 to about 1880, and then linear again, but with a smaller slope. This reflects a lower exponential growth rate after 1880.

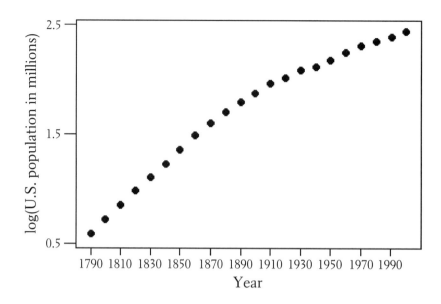

(c) We will use data from 1920 to 2000 since the pattern over that range appears very linear. Regressing log (POP) on YEAR yields the equation log $(\widehat{POP}) = -8.2761 + 0.005366$ (YEAR) with a correlation of $r = .998$. The predicted population in the year 2010 is 323,531,803.

(d) The residual plot (see below) shows random scatter, so the original data is exponential. The value of r^2 for the transformed data is 0.995.

Residuals Versus YEAR

(response is LOGPOP)

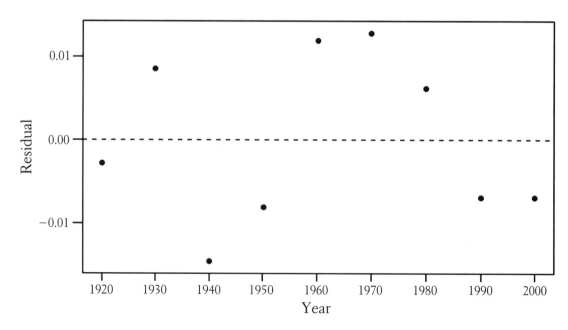

(e) The exponential model for these data is quite good.

4.12 (a) With the lengths of the fish in list LEN and the weights in list WT, logs of the lengths in list LOGL and the logs of the weights in list LOGW, and the power equation installed as Y_2, define list L_1 to be Y_2 (LEN). These are the predicted weights of the fish. Then define list L_2 to be the observed weights minus the predicted weights: WT-L_1. Then use the command STAT / 1-Var Stats L_2.

The sum of the squares of the deviations, $\Sigma x^2 = 7795.687$. Here are the lists:

Age	Len (cm)	Wt (g)	log (Len)	log (Wt)	Predict	Deviatns
1	5.2	2	0.71600	0.30103	1.923	0.0770
2	8.5	8	0.92942	0.90309	8.605	−0.6054
3	11.5	21	1.06070	1.32222	21.632	−0.6319
4	14.3	38	1.15534	1.57978	42.042	−4.0421
5	16.8	69	1.22531	1.83885	68.716	0.2835
6	19.2	117	1.28330	2.06819	103.253	13.7472
7	21.3	148	1.32838	2.17026	141.698	6.3020
8	23.3	190	1.36736	2.27875	186.302	3.6978
9	25.0	264	1.39794	2.42160	230.930	33.0696
10	26.7	293	1.42651	2.46687	282.232	10.7679
11	28.2	318	1.45025	2.50243	333.421	−15.4207
12	29.6	371	1.47129	2.56937	386.509	−15.5092
13	30.8	455	1.48855	2.65801	436.304	18.6958
14	32.0	504	1.50515	2.70243	490.238	13.7617
15	33.0	518	1.51851	2.71433	538.467	−20.4672
16	34.0	537	1.53148	2.72997	589.786	−52.7863
17	34.9	651	1.54283	2.81358	638.697	12.3032
18	36.4	719	1.56110	2.85673	726.149	−7.1486
19	37.1	726	1.56937	2.86094	769.576	−43.5762
20	37.7	810	1.57634	2.90848	808.162	1.8384

The quantity that was *minimized* was the sum of the squares of the deviations of the transformed points.

(b) Note that you can't use 1-Var Stats to find the sum of the squares of the residuals in the list RESID; you need to first copy the list RESID into a different list. The sum of the squares of the residuals is .0143.

(c) There's no reason to expect the answers to be the same.

4.13 Let x = weight of mammal in kg, y = lifespan of mammal in years. A plot of log y vs. log x for these data yields the following:

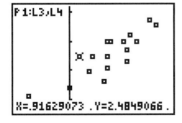
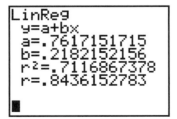

A power model with $a = 10^{0.7617} = 5.777$ and power $b = 0.2182$ appears to fit the data well. The model, $y = (10^{0.7617})x^{0.2182}$, predicts a lifespan of 17.06 years for humans. Oh well.

4.14 The heart weight (a 3-dimensional property) of these mammals should be proportional to the length (1-dimensional) of the cavity of the left ventricle. Here is a scatterplot of heart weight on length of left ventricle cavity.

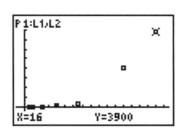

Although the scatterplot appears to be exponential, a power function model makes more sense in this setting. We therefore conjecture a model of the form WEIGHT = a × LENGTH^b. Plotting log(WEIGHT) vs. log(LENGTH), a linear pattern of points is obtained, confirming our choice of models.

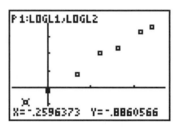

Next, linear regression is performed on the transformed points, and the LSRL is overlayed on the plot of the transformed points. The correlation is 0.997. The residual plot confirms that the power function model is appropriate.

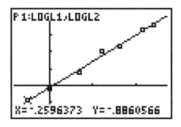

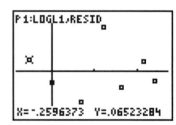

Performing the inverse transformation proceeds as follows:

$$\log \hat{y} = -.1364 + 3.1387 \log x = -.1364 + \log(x)^{3.1387}$$

$$10^{\log \hat{y}} = 10^{(-.1364 + \log(x) 3.1387)}$$

$$\hat{y} = (10^{-.1364})(x^{3.1387})$$

The power function plotted with the original data set (see facing page) shows a reasonably good model for this mammal heart data.

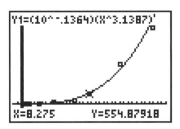

4.15 (a) As height increases, weight increases. Since weight is a 3-dimensional attribute and height is 1-dimensional, weight should be proportional to the cube of the height. One might speculate a model of the form WEIGHT = a × HEIGHT$^{\wedge b}$, where a and b are constants. This is a power function.

(b) Height is the explanatory variable, and weight is the response variable. Here is a scatterplot of the data.

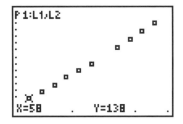

(c) Calculate the logarithms of the heights and the logarithms of the weights. Plot log(WT) vs. log(HT). The plot appears to be very linear, so least-squares regression is performed on the transformed data. The correlation is 0.99997. The regression line fits the transformed data extremely well.

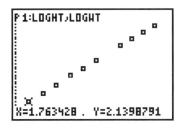
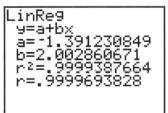
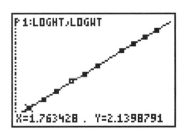

(d) Here is a residual plot for the transformed data.

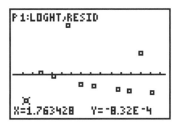

There is no discernable pattern. The line clearly fits the transformed data well.

(e) An inverse transformation yields a power equation for the original overweight data: $y = (10^{-1.39123})(x^{2.00286})$.

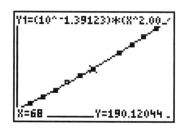

The predicted severely overweight value for a 5'10" adult (70 inches tall) would be $y = (10^{-1.39123})(70^{2.00286}) = 201.5$ pounds. The predicted severely overweight value for a 7 foot adult (84 inches tall) would be $y = (10^{-1.39123})(84^{2.00286}) = 290.3$ pounds.

4.16 (a) Below at left is a scatterplot of the pizza price data ($P1:L_1,L_2$). If price is proportional to surface *area* of the pizza, then power regression ($y = ax^b$) should be an appropriate model, where x = diameter of the pizza. On the TI-83, enter the logarithms of x in, say, list L_3, and the logs of y in list L_4. If a power function is an appropriate model, then the plot of log y on log x should be linear. Below at right is a plot of the transformed data ($P2:L_3, L_4$).

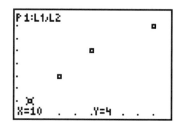

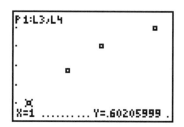

Performing least squares regression of log y vs. log x yields the following equation: log $\hat{y} = -.9172 + 1.15516$ log x. The correlation is $r = .976$. Performing the inverse transformation, we raise 10 to both sides of the equation and simplify to obtain $y = .121\ x^{1.5516}$. The last point (the 18" giant pizza) is out of line. Perhaps this size pizza is priced lower than it should be to give greater value. If this last point is deleted, as being nonrepresentative, and least-squares regression is performed on the remaining points, the power regression model becomes $y = 10^a x^b = .0348 x^{2.065}$. The power of x is very close to the 2 you would expect for an area, and the r value improves to 0.9988. This is the model we will adopt.

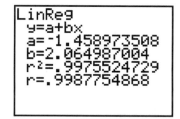

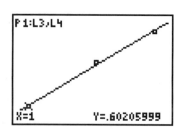

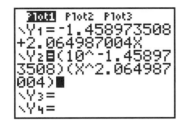 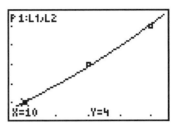

(Note that although a straight line and a logarithmic curve both fit these data better, we reject these models as inferior because of the area consideration.)

(b) Clearly, the giant pizza (18″) is underpriced. According to our model, the price of an 18″ pizza should cost $Y_2(18) = .0348(18)^{2.065} = 13.59$.

(c) A new 6″ pizza should cost $1.95 under the first power function model and $1.41 according to the second power model.

(d) A 24″ pizza would cost $16.76 under the first model and $24.61 under the second model.

4.17 (a) $y = 500(1.075)^{year}$. 537.50, 577.81, 621.15, 667.73, 717.81, 771.65, 829.52, 891.74, 958.62, 1030.52.

(b)

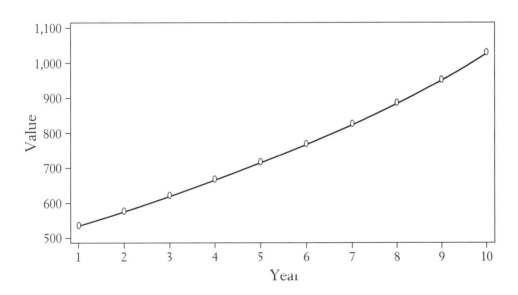

(c) The logs are 2.73, 2.76, 2.79, 2.82, 2.86, 2.89, 2.92, 2.95, 2.98, 3.01.

4.18 Alice has $500(1.075)^{25} = 3049.17$, Fred has $500 + 100(25) = 3000.00$.

4.19 (a) See scatterplot below. The association is positive; as the length of the fish increases, its weight increases. The association is strong, and the pattern is clearly curved.

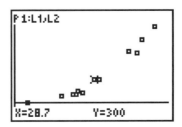

(b) Considering that length is one-dimensional and weight is three-dimensional, we would expect the weight of the fish to be proportional to the cube of its length. This is a power law relationship, so using logarithms to transform both length and weight should straighten the scatterplot. A plot of log (WEIGHT) vs. log (LENGTH) (see below) shows a very linear pattern. The correlation of log (WEIGHT) and log (LENGTH) is $r = 0.994$ ($r^2 = 0.9886$).

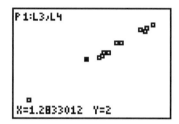 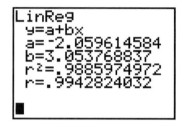

4.20 The scatterplot of width against length (see below) is considerably more linear than the plot of weight against length in Exercise 4.19. This is not surprising because width and length are both one-dimensional.

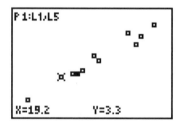

4.21 (a) See graphs below. (b) From 0 to 6 hours (the first phase), the growth is "flat"—the colony size does not change much. From 6 to 24 hours, the log(mean colony size) plot looks like a positively sloped line, suggesting a period of exponential growth. At some point between 24 and 36 hours, the growth rate drops off (the 36-hour point is considerably below the linear pattern of the 6- to-24-hour points). (c) The regression equation is $\hat{y} = -0.594 + 0.0851x$; the prediction for $x = 10$ hours is $\hat{y} \doteq 0.257$, so the predicted mean colony size is about $10^{0.257} = 1.81$.

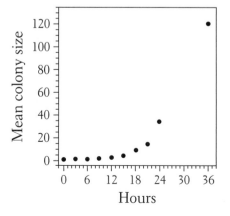 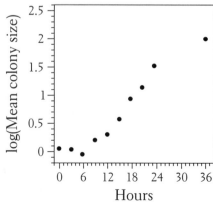

4.22 (a) We are given $\ln y = -2 + 2.42 \ln x$. Therefore, using the laws of logarithms, we have:

$$e^{\ln y} = e^{-2 + 2.42 \ln x}$$

$$y = e^{-2} \times e^{2.42 \ln x} = e^{-2} \times e^{\ln x^{2.42}} = e^{-2} \times x^{2.42}$$

(b) At $x = 30$, $y = e^{-2} \times (30)^{2.42} \approx 508.21$.

4.23 For hours 6–24 and log (mean colony size), $r = 0.9915$. For time and log (individual colony size). $r^* = 0.9846$. This is smaller because individual measurements have more scatter (see scatterplot, below): the points do not cluster as tightly around a line.

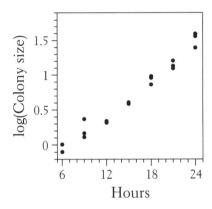

4.24 (a) The plot of distance against height (see below) resembles a power curve x^p with $p < 1$. Since we want to pull in the right tail of the distribution, we should apply a transformation x^p with $p < 1$.

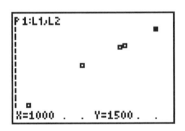

(b) A plot of distance against the *square root* of height (see below) straightens the graph quite nicely.

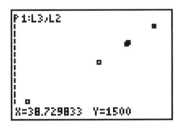

4.25 (a) See scatterplot below. The association is negative, nonlinear, and strong. Species with seeds that weigh the least tend to produce the largest count of seeds. Species with the heaviest seeds produce the smallest number of seeds.

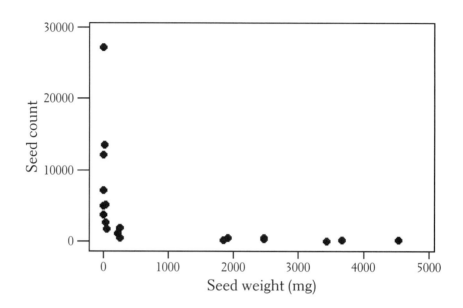

(b) A plot of log (COUNT) on log (WEIGHT) (see below) shows a moderately strong linear pattern with correlation $r = -0.93$ and $r^2 = 0.86$. The transformed data show a negative association.

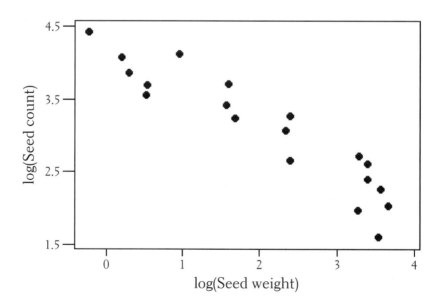

4.26 (a) Answers vary.

(b) The population of cancer cells at the end of Year $n - 1$ is easily seen, by the indicated pattern, to be $P = P_0 (7/6)^{n-1}$. The population of cancer cells at the end of Year n is then $P = P_0 (7/6)^{n-1} + (1/6)(P_0(7/6)^{n-1}) = P_0 (7/6)^n$.

(c) Answers vary.

4.27 (a) Regression line: $\hat{y} = 1166.93 - 0.58679x$. (b) Based on the slope, the farm population decreased about 590 thousand (0.59 million) people per year. The regression line explains 97.7% of the variation. (c) $-782,100$ — clearly a ridiculous answer, since a population must be greater than or equal to 0.

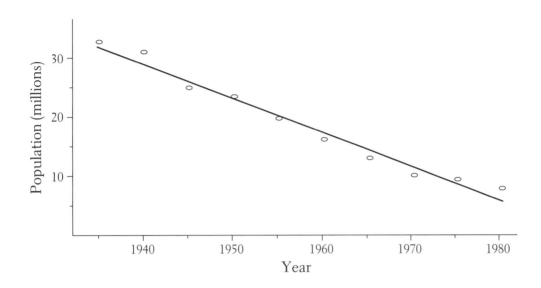

4.28 The explanatory and response variables were "consumption of herbal tea" and "cheerfulness." The most important lurking variable is social interaction—many of the nursing home residents may have been lonely before the students started visiting.

4.29 The correlation would be smaller because there is much more variation among the individual data points. This variation could not be as fully explained by the linear relationship between speed and step rate.

4.30 No; more likely it means that patients with more serious conditions (which require longer stays) tend to go to larger hospitals, which are more likely to have the facilities to treat those problems.

4.31 The correlation would be lower; individual stock performances will be more variable, weakening the relationship.

4.32 (a) Scatterplot is below. The influential observation (circled) is observation 7, (105, 89).

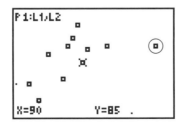

(b) In the plot on the next page, the line $\hat{y} = 20.49 + 0.754x$ (that is, the line with the larger slope) is the line that omits the influential observation (105, 89). The line $\hat{y} = 50.01 + 0.410x$ is the line closer to this point. Influential observations "attract" the regression line.

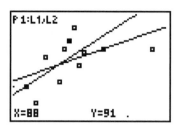

4.33 Seriousness of the fire is a lurking (common–response) variable: more serious fires require more attention and do more damage. It would be more accurate to say that a large fire "causes" more firefighters to be sent, rather than vice versa.

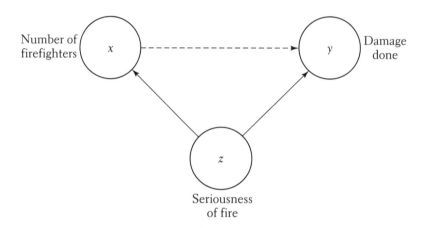

4.34 A reasonable explanation is that the cause-and-effect relationship goes in the other direction: Doing well makes students feel good about themselves, rather than vice versa.

4.35 (a) $r^2 = 0.925$ —more than 90% of the variation in one SAT score can be explained through a linear relationship with the other score. (b) The correlation would be much smaller, since individual students have much more variation between their scores. Some may have greater verbal skills and low scores in math (or vice versa); some will be strong in both areas, and some will be weak in both areas. By averaging—or, as in this case, taking the median of—the scores of large groups of students, we muffle the effects of these individual variations.

4.36 Age is the lurking (common–response) variable here: we would expect both quantities—shoe size and reading level—to increase as a child ages.

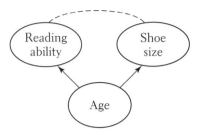

4.37 The explanatory variable is whether or not a student has taken at least two years of foreign language, and the score on the test is the response. The lurking variable is the students' English skills *before* taking (or not taking) the foreign language: students who have a good command of English early in their high school career are more likely to choose (or be advised to choose) to take a foreign language and to do well on an English test in *any* case.

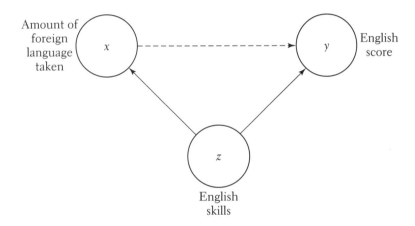

4.38 In this case, there may be a causative effect, but in the direction opposite to the one suggested: People who are overweight are more likely to be on diets, and so choose artificial sweeteners over sugar. (Also, heavier people are at a higher risk to develop diabetes; if they do, they are likely to switch to artificial sweeteners.)

4.39 Time standing up is a confounding variable in this case. The diagram below illustrates the confounding between exposure to chemicals and standing up.

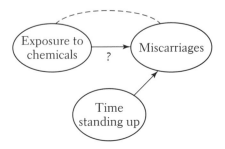

4.40 A student's intelligence may be a lurking (confounding) variable: Stronger students (who are more likely to succeed once they get to college) are more likely to choose to take these math courses, while weaker students may avoid them. Other possible answers might be variations on this idea; for example, if we believe that success in college depends on a student's self-confidence, we might suppose that confident students are more likely to choose math courses.

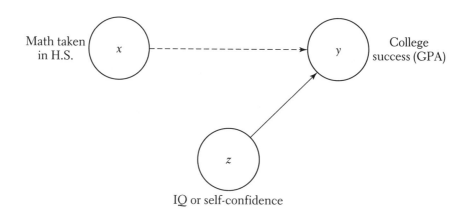

4.41 This is a case of common response. It could be that children with lower intelligence watch many hours of TV *and* get lower grades as well. Or children from lower socioeconomic households where the parent(s) are less likely to limit television viewing and be unable to help their children with their schoolwork because the parents themselves lack education may watch more TV *and* get lower grades as a result.

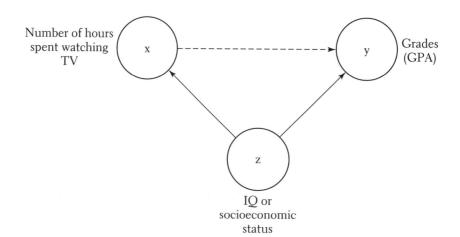

4.42 Several possible lurking variables could help explain this supposed "causal" relationship. Factors such as innate intelligence and experience in English and mathematics classes are hopelessly confounded with the effects of music experience on test scores. Both music experience and test scores could also be responding to the socioeconomic status of the student. Students from affluent families may be more likely to receive extensive exposure to music and may also do better on tests because they can afford test-taking courses and other costly means of preparation.

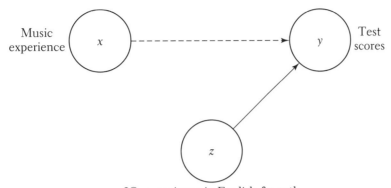

IQ, experience in English & math

4.43 The effects of coaching are confounded with those of experience. A student who has taken the SAT once may improve his or her score on the second attempt because of increased familiarity with the test.

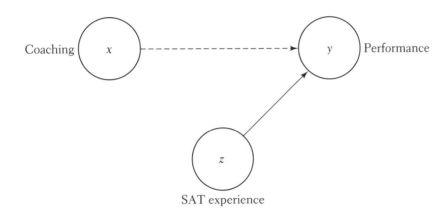

SAT experience

4.44 The plot below is a very simplified (and not very realistic) example—open circles are economists in business; filled circles are teaching economists. The plot should show positive association when either set of circles is viewed separately, and should show a large number of bachelor's degree economists in business and graduate degree economists in academia.

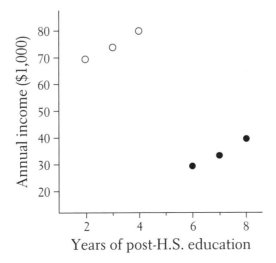

4.45 Spending more time watching TV means that *less* time is spent on other activities: these may suggest lurking (confounding) variables. For example, perhaps the parents of heavy TV watchers do not spend as much time at home as other parents. Also, heavy TV watchers would typically not get as much exercise. Another factor could be the economy; as the economy has grown over the past 20 years, more families can afford TV sets (even multiple TV sets), and as a result, TV viewing has increased *and* children have less physical work to do in order to make ends meet.

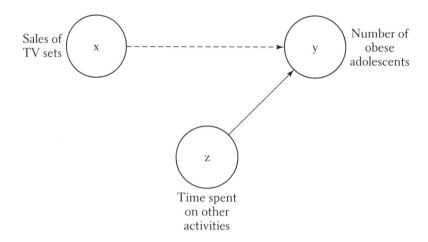

4.46 The correlation would be lower: individual stock performances would be more variable, weakening the relationship.

4.47 Higher income can cause better health: higher income means more money to pay for medical care, drugs, and better nutrition, which in turn results in better health. Better health can cause higher income: if workers enjoy better health, they are better able to get and hold a job, which can increase their income.

4.48 (a) The plot below includes the point with the largest residual (18.6, 69.4).

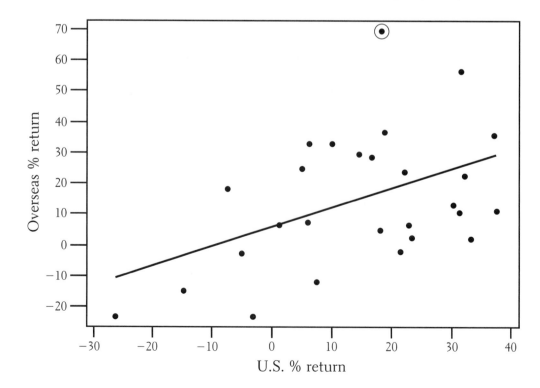

The plot below deletes (18.6, 69.4). The deleted point does not appear to be very influential.

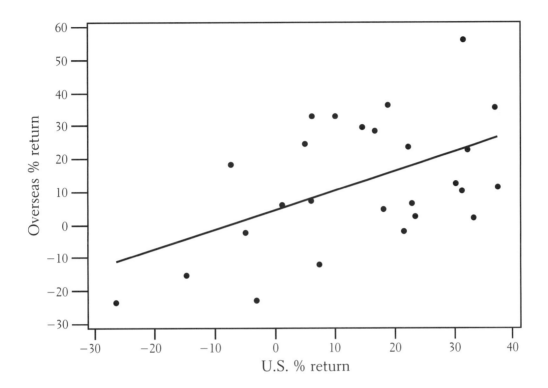

(b) The residual plot (see below) does not appear to have any clear pattern, though the residuals at the extreme right appear a little more "negative" than comparable residuals at the left.

Residuals versus the order of the data

(response is OVERSEAS)

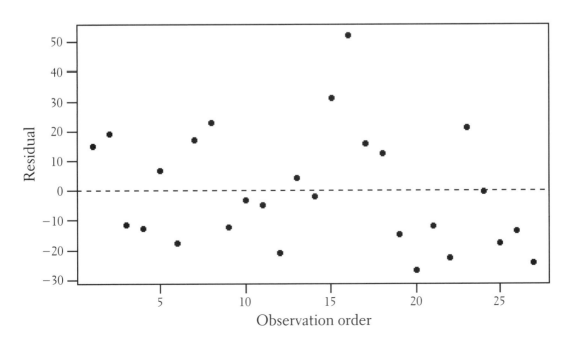

4.49 (a) Yes. The relationship is not very strong. (b) In hospitals that treat very few heart attack cases, the mortality is quite high. One can almost see an exponentially decreasing pattern of points in the plot. The nonlinearity weakens the conclusion that heart attack patients should avoid hospitals that treat few heart attacks. The mortality rate is extremely variable for those hospitals that treat few heart attacks.

4.50 The sum is 56,007. This sum differs from the listed column sum of 56,008 because of round-off error. (We are measuring in thousands of persons.)

4.51 21.6%, 46.5%, and 32.0% (total is 100.1% due to rounding).

4.52 11.8%, 11.2%, and 25.4%. The percentage of people who did not finish high school is about the same for the 25–34 and 35–54 age groups, but more than doubles for the 55 and over age group.

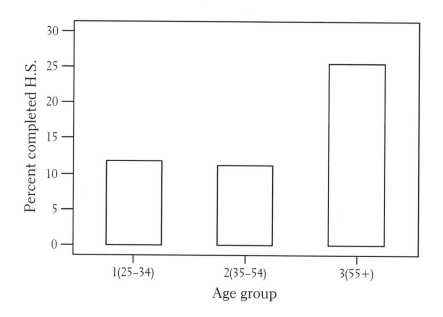

4.53 (a) 5375. (b) 1004/5375 = 18.7%. (c) Both parents smoke: 1780 (33.1%); one parent smokes: 2239 (41.7%); neither parent smokes: 1356 (25.2%).

4.54 (a) Below. (b) Cold water: $\frac{16}{27} \doteq 59.3\%$; Neutral: $\frac{38}{56} \doteq 67.9\%$; Hot water: $\frac{75}{104} \doteq 72.1\%$. The percentage hatching increases with temperature; the cold water did not prevent hatching, but made it less likely.

	Temperature		
	Cold	Neutral	Hot
Hatched	16	38	75
Did not hatch	11	18	29
Total	27	56	104

4.55 (a) 6014; 1.26%. (b) Blood pressure is explanatory. (c) Yes: among those with low blood pressure, 0.785% died; the death rate in the high blood pressure group was 1.65%—about twice as high as the other group.

4.56 Compute: $14424/56008 = 25.4\%$, $20060/56008 = 35.8\%$, etc.

4.57 Among 35- to 54-year-olds: 11.2% never finished high school, 32.5% finished high school, 27.8% had some college, and 28.5% completed college. This is more like the 25–34 age group than the 55 and over group.

4.58 Among those with 4 or more years of college: 24.5% are 25–34, 51.7% are 35–54, and 23.6% are 55 or older.

4.59 (a) Use column percentages (e.g., $\frac{68}{225} \doteq 30.2\%$ of females are in administration, etc.). See the table below. The biggest difference between women and men is in Administration: a higher percentage of women chose this major. Meanwhile, a greater proportion of men chose other fields, especially finance. (b) There were 386 responses; $\frac{336}{722} \doteq 46.5\%$ did not respond.

	Female	Male
Accting.	30.2%	34.8%
Admin.	40.4%	24.8%
Econ.	2.2%	3.7%
Fin.	27.1%	36.6%

4.60 (a) Below. (b) 70% of male applicants are admitted, while only 56% of females are admitted. (c) 80% of male business school applicants are admitted, compared with 90% of females; in the law school, 10% of males are admitted, compared with 33.3% of females. (d) Six out of 7 men apply to the business school, which admits 83% of all applicants, while 3 of 5 women apply to the law school, which admits only 27.5% of its applicants.

	Admit	Deny
Male	490	210
Female	280	220

4.61 (a) Below. (b) Overall, 11.9% of white defendants and 10.2% of black defendants get the death penalty. However, for white victims, the percentages are 12.6% and 17.5% (respectively); when the victim is black, they are 0% and 5.8%. (c) In cases involving white victims, 14% of defendants got the death penalty; when the victim was black, only 5.4% of defendants were sentenced to death. White defendants killed whites 94.3% of the time—but are less likely to get the death penalty than blacks who killed whites.

	Yes	No
White defendant	19	141
Black defendant	17	149

4.62 (a) Adding across the bottom (total) row: 14,116 thousand, or 14,116,000. (Adding across the rows, then down the columns gives 14,117 thousand.) (b) Add across the second row of the table; this appears at the right end of the second row of the table: $\frac{7771}{14,116} \doteq 55.1\%$. (Using 14,117, this rounds to 55.0%.) (c) Reading across the second entry of the second row of the table: $\frac{1378}{1966} \doteq 70.1\%$, $\frac{1198}{3473} \doteq 34.5\%$, $\frac{4607}{6119} \doteq 75.3\%$, $\frac{588}{2559} \doteq 23.0\%$. (d) 18- to 24-year-olds constitute the majority of full-time students at both 2- and 4-year institutions, but make up a much smaller proportion of part-time students.

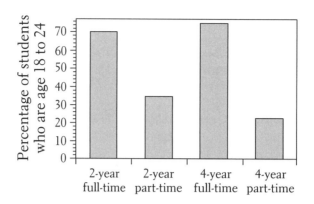

4.63 (a) There are 3472 thousand 2-year part-time students; $41.1\% \doteq \frac{1427}{3472}$ are 25 to 39 years old. (b) There are 4388 thousand 25- to 39-year-old students; $32.5\% \doteq \frac{1427}{4388}$ are enrolled part-time at 2-year colleges.

4.64 (a) These are in the right-hand "margin" of the table: Adding across the rows, we find 286 (thousand), 7771, 4388, and 1672, respectively. Dividing by 14,116 (or 14,177) gives 2.0%, 55.0% (or 55.1%), 31.1%, and 11.8%.

(b) From the "2-year part-time" column, we divide 125, 1198, 1427, and 723 by 3473 to get 3.6%, 34.5%, 41.1%, and 20.8%.

(c) Two-year part-time students are more likely to be older (over 24, and even more so over 40) than undergraduates in general. They are also slightly more likely to be under 18, and considerably less likely to be 18 to 21.

(d) Roundoff error accounts for the difference (3473 versus 3472).

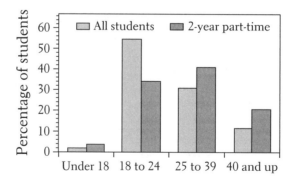

4.65 For older students, take 119, 723, 225, and 605, and divide by 1672 to get 7.1%, 43.2%, 13.5%, and 36.2%. We might then compare these with the same percentages for the whole population: From the bottom row of the table, divide 1966, 3473, 6119, and 2559 by 14,116 to get 13.9%, 24.6%, 43.3%, and 18.1%.

From these percentages, and the bar chart on the facing page, we can see that older students are more likely to be part-time than students in general.

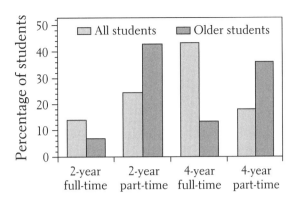

4.66 (a) At the right end of the second row of the table, $\frac{7771}{14,116} \doteq 55.1\%$. (Using 14,117, this rounds to 55.0%.)

(b) $\frac{1378 + 1198}{1966 + 3472} \doteq 47.4\%$.

(c) $\frac{1198 + 588}{3472 + 2559} \doteq 29.6\%$.

4.67 After adding the shotgun and rifle numbers, two approaches are possible: (1) Compute the conditional distribution of homicides versus suicides for each gun type (that is, view gun type as the explanatory variable). For handguns, we find 79.1% of handgun uses were homicides, while only 20.9% were suicides. Meanwhile, long gun uses were almost evenly split: 48.3% homicides, 51.7% suicides. The bar graph on the left below shows this comparison. (2) Compute the conditional distribution of gun type for homicides versus suicides (that is, view crime type as explanatory). Among homicides, 89.3% were committed with handguns, 8.2% with long guns, and 2.5% unknown. Among suicides, 70.9% used handguns, 26.3% long guns, and 2.9% unknown (that is, long guns were more often used in suicides). The bar graph on the right below shows this comparison.

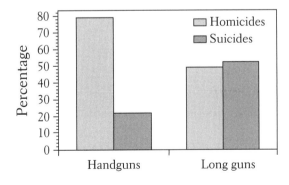
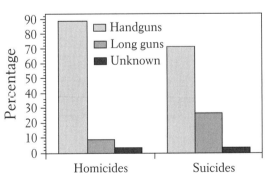

4.68 (a) 58.3% of desipramine users did not have a relapse, while 25.0% of lithium users and 16.7% of those who received placebos succeeded in breaking their addictions. Note that use of percentages is not as crucial here as in other cases, since an equal number of drug addicts was assigned to each drug.

(b) Because random assignment was used, we can *tentatively* assume causation (though there are other questions we need to consider before we can reach that conclusion).

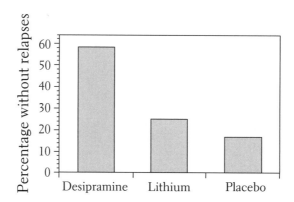

4.69 (a) 7.2%. (b) 10.1% of the restrained children were injured, compared to 15.4% of unrestrained children.

4.70 (a) See below. (b) Joe: .240, Moe: .260. Moe has the best overall batting average. (c) Against right-handed pitchers: Joe: .400, Moe: .300. Against left-handed pitchers: Joe: .200, Moe: .100. Joe is better against both kinds of pitchers. (d) Both players do better against right-handed pitchers than against left-handed pitchers. Joe spent 80% of his at-bats facing left-handers, while Moe only faced left-handers 20% of the time.

	Hit	No hit
All pitchers		
Joe	120	380
Moe	130	370
Right-handed		
Joe	40	60
Moe	120	280
Left-handed		
Joe	80	320
Moe	10	90

4.71 Examples will vary, of course; here is one very simplistic possibility (the two-way table is immediately below; the three-way tables are below the two-way table). The key is to be sure that the three-way table has a lower percentage of overweight people among the smokers than among the nonsmokers.

	Early death	
	Yes	No
Overweight	4	6
Not overweight	5	5

Smoker	Early death	
	Yes	No
Overweight	1	0
Not overweight	4	2

Nonsmoker	Early death	
	Yes	No
Overweight	3	6
Not overweight	1	3

4.72 (a)

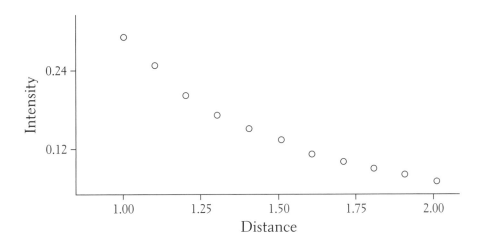

The data appear to follow a model of the form $y = ax^b$ where b is some negative number. Taking logs of both sides of this model equation, we have

$$\log y = \log a + \log x^b$$
$$= \log a + \log b \, (\log x)$$

We perform linear regression on the $\log x$ and $\log y$ data and obtain the following least-squares regression model:

$$\log y = -.5235 - 2.013 \log x$$

The inverse transformation yields

$$y = (10^{-.5235})(x^{-2.013})$$
$$= -3 \, x^{-2.013}$$

(b) $y = 0.3 \, x^{-2.013}$. (c) The intensity of the light bulb appears to vary inversely with (approximately) the square of the distance from the bulb. (d) The formula for intensity as a function of distance is given by $y = 900/x^2$ where y is measured in *candlepower*, x in meters. With an appropriate units change to candelas, this formula appears to fit the experimental data.

4.73 (a) The scatterplot (P1) of period vs. length of pendulum suggests a power function model. The plot (P2) of log (period) vs. log (length) appears linear. There is a positive association; as length increases, the period increases.

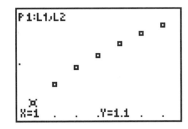
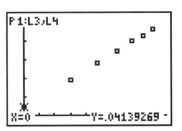

(b) Least-squares regression is performed on the transformed data. The correlation between the transformed variables is extremely strong. The LSRL fits the transformed data like a glove.

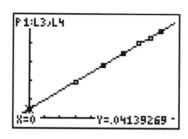 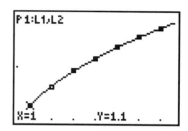

The power function model takes the form $\hat{y} = (10^{.0415})(x^{.5043})$.

(c) The equation for the model says that the period of a pendulum is proportional to the square root of its length.

4.74 (a) 1, 2, 4, 8, 16, 32, 64, 128, 256, 512.

(b)

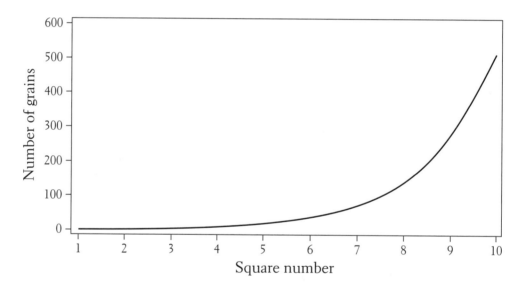

(c) Approximately 9,000,000,000,000,000,000.

(d) 0.00, 0.30, 0.60, 0.90, 1.20, 1.51, 1.81, 2.11, 2.41, 2.71.

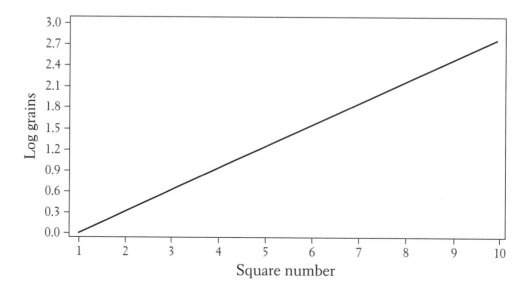

(e) $b = .3$, $a = -.3$, $-.3 + .3(64) = 18.9$. The log of the answer in part (c) is 18.95.

4.75 (a) Plotting the logarithm of men's times against year and regressing, we find that the regression equation is $\log \hat{y} = 3.141 - .00057x$, which translates to an exponential model of $\hat{y} = (10^{3.141})(10^{-.00057})^x = 1383.6(.9987)^x$. This model does not fit the data very well.

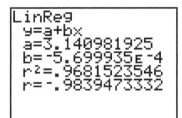
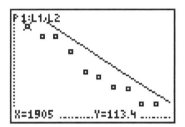

Plotting the log of men's times against the log of year and regressing, we find that the regression equation is $\log \hat{y} = 10.45 - 2.56 \log x$, which translates to a power model of $\hat{y} = (10^{10.45})x^{-2.56}$. This model provides a much better fit (note that r is closer to -1).

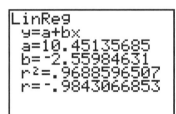
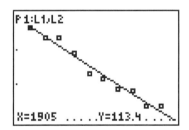

(b) Plotting the logarithm of women's times against year and regressing, we find that the regression equation is $\log \hat{y} = 5.12 - .0015x$, which translates to an exponential model of $\hat{y} = (10^{5.12})(10^{-.0015})^x$. This model fits the data well.

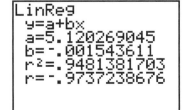
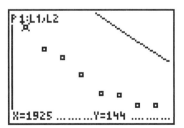

Plotting the log of women's times against the log of year and regressing, we find that the regression equation is $\log \hat{y} = 25.051 - 6.973 \log x$, which translates to a power model of $\hat{y} = (10^{25.051})x^{-6.973}$. This model provides a slightly better fit (as in (a), the value of r is slightly closer to -1).

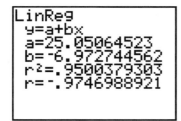

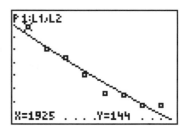

In both the men's and the women's cases, the power model is a slightly better model for the data.

(c) The power curves in (a) and (b) will never reach 0 because the horizontal axis serves as an asymptote for these curves; the curves approach the axis but never touch it. The curves intersect in approximately the year 2035, corresponding to a time of 95.477 seconds.

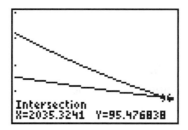

(d) The power models are reasonably accurate. However, they may not take into account the fact that, as we move forward in time, the rates at which the curves tend to "flatten out" may themselves change. We are still not convinced that the women's record will ultimately "catch up" to the men's record.

4.76 (a) The pattern appears to be exponential.

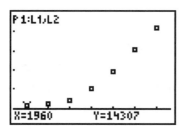

(b) The plot of log SPENDING vs. YEAR appears linear, so the exponential growth model fits well.

(c) See plot at right below.

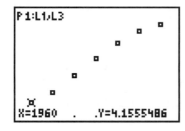

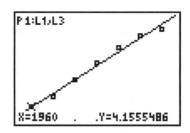

(d) Predicted log for 1988 is 5.617046, predicted amount spent in 1988 is $414,043.74.

(e) Log is 5.55. The point is somewhat below the line. The rate of growth appears to be slower.

4.77 These data call for an exponential decay model. Plotting log(count) versus time (see below) produces a very linear pattern. Regressing the reexpressed data gives the regression equation log(count) = 2.5941 − .0949 (minutes) and correlation −0.9942. A residual plot from the regression shows no clear pattern, so an exponential equation will fit the original data well.

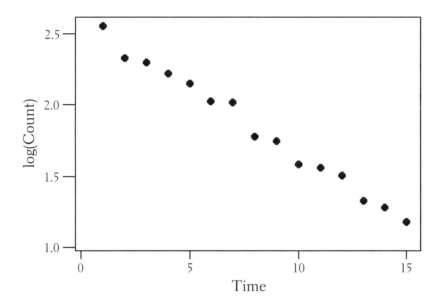

Residuals versus the order of the data
(response is LOGCOUNT)

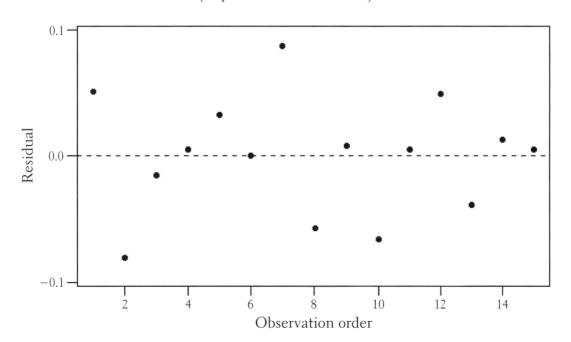

4.78 A scatterplot of EFT against year (see below) reveals a clearly nonlinear growth pattern. From 1985 to 1996, the growth pattern appears to be exponential, with a fairly consistent rate of growth. After 1996, the growth pattern remains exponential, but the rate of growth begins to slow down. In particular, there is very little growth from 1998 to 1999, possibly indicating that the number of EFTs was pressing against current technological limitations during this period.

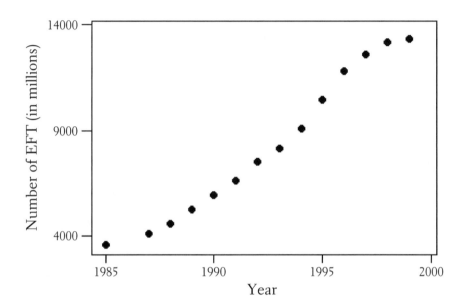

If log(EFT) is plotted against year for the years 1985–1996 (see below), the plot is very nearly linear. The regression equation for these data is log (EFT) = −92.4 + 0.0483 YEAR, with r^2 = 0.996.

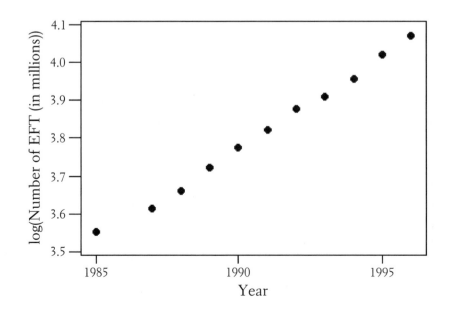

4.79

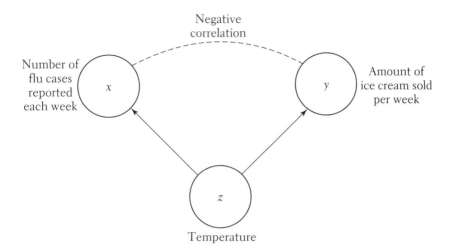

Both x and y are responding to the variable z = *temperature*. In the winter, when temperatures are low, there are many flu cases (x) but relatively little ice cream sold (y). In the summer, when temperatures are high, there are few cases of flu, but a large amount of ice cream is sold. This is an example of *common response*.

4.80 (a) The percentages of people who voted were: 62.8%; 61.9%; 60.9%; 55.2%; 53.5%; 52.8%; 53.3%; 50.3%; 55.1%; 49.1%; and 50.4%. The plot of % voted vs. year is given below.

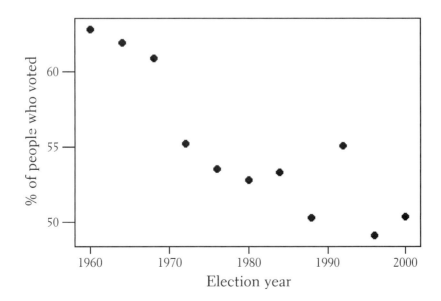

(b) Since young people are less likely to vote, the lowering of the minimum voting age in 1970 brought a large number of individuals with a low turnout into the population of eligible voters, thereby driving down the voting percentage.

4.81 (a) 19,039; roundoff error. (b) 25.1%, 54.7%, 10.0%, and 10.2%. See bar graph below. (c) 15–24: 84.7%, 14.1%, 0.1%, 1.1%; 40–64: 8.4%, 70.1%, 5.5%, 16.0%. Among the younger women, more than 4 out of 5 have not yet married, and those who are married have had little time to become widowed or divorced. Most of the older group is or has been married—only about 8.4% are still single. (d) 58.1% of single women are 15–24, 26.7% are 25–39, 12.8% are 40–64, and 2.4% are 65 or older.

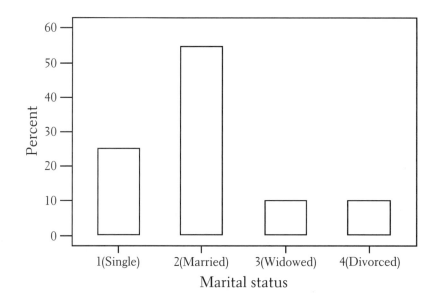

4.82 Apparently women are more likely to be in fields that pay less overall (to both men and women). For example, if many women, and few men, have Job A, where they earn $40,000 per year, and meanwhile few women and many men have Job B earning $50,000 per year, then lumping all women and men together leads to an incorrect perception of unfairness

4.83 (a) Adding corresponding entries in each table gives the table below. 76.1% of the smokers stayed alive for 20 years; 68.6% of the nonsmokers did so.

	Smoker	Not	Total
Dead	139	230	369
Alive	443	502	945
Total	582	732	1314

(b) For the youngest group, 96.2% of the nonsmokers and 93.4% of the smokers were alive 20 years later. For the middle group, 73.9% of the nonsmokers and 68.2% of the smokers were alive 20 years later. For the oldest group, 14.5% of the nonsmokers and 14.3% of the smokers were alive 20 years later. So the results are reversed when the data for the three age groups are combined.

(c) 45.9% of the youngest group, 55.2% of the middle aged group, but only 20.2% of the oldest group were smokers.

Producing Data

5.1 The population is employed adult women, the sample is the 48 club members who returned the survey.

5.2 (a) An individual is a person; the population is all adult U.S. residents. (b) An individual is a household; the population is all U.S. households. (c) An individual is a voltage regulator; the population is all the regulators in the last shipment.

5.3 This is an experiment: A treatment is imposed. The explanatory variable is the teaching method (computer assisted or standard), and the response variable is the increase in reading ability based on the pre- and posttests.

5.4 We can never know how much of the change in attitudes was due to the explanatory variable (reading propaganda) and how much to the historical events of that time. The data give no information about the effect of reading propaganda.

5.5 Observational. The researcher did not attempt to change the amount that people drank. The explanatory variable is alcohol consumption. The response variable is survival after 4 years.

5.6 (a) The data was collected after the anesthesia was administered. (b) Type of surgery, patient allergy to certain anesthetics, how healthy the patient was before the surgery.

5.7 Only persons with a strong opinion on the subject—strong enough that they are willing to spend the time, and 50 cents—will respond to this advertisement.

5.8 Letters to legislators are an example of a voluntary response sample—the proportion of letters opposed to the insurance should not be assumed to be a fair representation of the attitudes of the congresswoman's constituents.

5.9 Labeling from 001 to 440, we select 400, 077, 172, 417, 350, 131, 211, 273, 208, and 074.

5.10 Starting with 01 and numbering down the columns, one chooses 04-Bonds, 10-Fleming, 17-Liao, 19-Naber, 12-Goel, and 13-Gomez.

5.11 Assign 01 to 30 to the students (in alphabetical order). The exact selection will depend on the starting line chosen in Table B; starting on line 123 gives 08-Ghosh, 15-Jones, 07-Fisher, and 27-Shaw. Assigning 0–9 to the faculty members gives (from line 109) 3-Gupta and 6-Moore. (We could also number faculty from 01 to 10, but this requires looking up 2-digit numbers.)

5.12 Label the 500 midsize accounts from 001 to 500, and the 4400 small accounts from 0001 to 4400. We first encounter numbers 417, 494, 322, 247, and 097 for the midsize group, then 3698, 1452, 2605, 2480, and 3716 for the small group.

5.13 (a) Households without telephones, or with unlisted numbers. Such households would likely be made up of poor individuals (who cannot afford a phone), those who choose not to have phones, and those who do not wish to have their phone number published.

(b) Those with unlisted numbers would be included in the sampling frame when a random-digit dialer is used.

5.14 The higher no-answer was probably the second period—more families are likely to be gone for vacations, etc. Nonresponse of this type might underrepresent those who are more affluent (and are able to travel).

5.15 The first wording would pull respondents toward a tax cut because the second wording mentions several popular alternative uses for tax money.

5.16 *Variable*: Approval of president's job performance. *Population*: Adult citizens of the U.S., or perhaps just registered voters. *Sample*: The 1210 adults interviewed. *Possible sources of bias*: Only adults with phones were contacted. Alaska and Hawaii were omitted.

5.17 (a) 13,147 + 15,182 + 1448 = 29,777. (b) There's nothing to prevent a person from answering several times. Also, the respondents were only those who went to that Web site and took the time to respond. We cannot define "nonresponse" in this situation. (c) The results are slanted toward the opinions of men, who might be less likely to feel that female athletes should earn as much as men.

5.18 (a) The wording is clear. The question is somewhat slanted in favor of warning labels. (b) The question is clear, but it is clearly slanted in favor of national health insurance by asserting it would reduce administrative costs. (c) The question could be clearer by using simpler language. It is slanted in favor of incentives by starting out discussing environmental degradation.

5.19 (a) The adults in the country. (b) All the wood sent by the supplier. (c) All households in the U.S.

5.20 The call-in poll is faulty in part because it is a voluntary sample. Furthermore, even a small charge like 50 cents can discourage some people from calling in—especially poor people. Reagan's Republican policies appealed to upper-class voters, who would be less concerned about a 50 cent charge than lower-class voters who might favor Carter.

5.21 Number the bottles across the rows from 01 to 25, then select 12–B0986, 04–A1101, and 11–A2220. (If numbering is done down columns instead, the sample will be A1117, B1102, and A1098.)

5.22 In order to increase the accuracy of its poll results. Larger samples give more accurate results than smaller samples.

5.23 The blocks are already marked; select three-digit numbers and ignore those that do not appear on the map. This gives 214, 313, 409, 306, and 511.

5.24 (a) False—if it were true, then after looking at 39 digits, we would know whether or not the 40th digit was a 0, contrary to property 2. (b) True—there are 100 pairs of digits 00 through 99, and all are equally likely. (c) False—0000 is just as likely as any other string of four digits.

5.25 It is *not* an SRS, because some samples of size 250 have no chance of being selected (e.g., a sample containing 250 women).

5.26 (a) This question will likely elicit more responses against gun control (that is, more people will choose 2). The two options presented are too extreme; no middle position on gun control is allowed.

(b) The phrasing of this question will tend to make people respond in favor of a nuclear freeze. Only one side of the issue is presented.

(c) The wording is too technical for many people to understand—and for those that *do* understand it, it is slanted because it suggests reasons why one should support recycling. It could be rewritten to something like: "Do you support economic incentives to promote recycling?"

5.27 A smaller sample gives less information about the population. "Men" constituted only about one-third of our sample, so we know less about that group than we know about all adults.

5.28 The chance of being interviewed is 3/30 for students over age 21 and 2/20 for students under age 21. This is 1/10 in both cases. It is not an SRS because not all combinations of students have an equal chance of being interviewed. For instance, groups of 5 students all over age 21 have no chance of being interviewed.

5.29 Answers will vary, of course. One possible approach: Obtain a list of schools, stratified by size or location (rural, suburban, urban). Choose SRSs (not necessarily all the same size) of schools from each strata. Then choose SRSs (again, not necessarily the same size) of students from the selected schools.

5.30 (a) Split the 200 addresses into 5 groups of 40 each. Looking for 2-digit numbers from 01 to 40, we find 35, and so take 35, 75, 115, 155, and 195. (b) Every address has a 1-in-40 chance of being selected, *but* not every subset has an equal chance of being picked—for example, 01, 02, 03, 04, and 05 cannot be selected by this method.

5.31 Units are the individual trees. Factor is the amount of light. Treatments are full light and reduced light. Response variable is the weight of the trees.

5.32 The liners are the experimental units. The heat applied to the liners is the factor; the levels are 250°F, 275°F, 300°F, and 325°F. The force required to open the package is the response variable.

5.33 The units are the individuals who were called. One factor is what information is offered. Treatments are (1) giving name, (2) identifying university, (3) both of these. Second factor is offering to send a copy of the results. The treatments are either offering or not offering. The response is whether the interview was completed.

5.34 Subjects: 300 sickle cell patients. Factor: drug given. Treatments: hydroxyurea and placebo. Response variable: number of pain episodes.

5.35 (a) This is an experiment, since the teacher imposes treatments (instruction methods). (b) The explanatory variable is the method used (computer software or standard curriculum), and the response is the change in reading ability.

5.36 (a) The experimental units are the batches of the product; the yield of each batch is the response variable. (b) There are two factors: temperature (with 2 levels) and stirring rates (with 3 levels), for a total of 6 treatments. (c) Since two experimental units will be used for each treatment, we need 12.

| | | Factor B: Stirring rates | | |
		60 rpm	90 rpm	120 rpm
Factor A:	50°C	1	2	3
Temperature	60°C	4	5	6

5.37 (a) In a serious case, when the patient has little chance of surviving, a doctor might choose not to recommend surgery; it might be seen as an unnecessary measure, bringing expense and a hospital stay with little benefit to the patient.

(b)

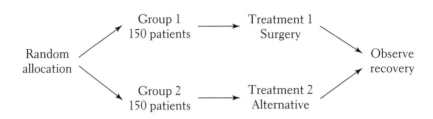

5.38 (a)

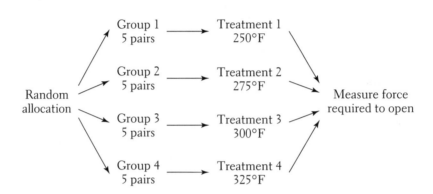

(b) Number the liners from 01 to 20, then take Group 1 to be 16, 04, 19, 07, and 10; Group 2 is 13, 15, 05, 09, and 08; Group 3 is 18, 03, 01, 06, and 11. The others are in Group 4.

5.39 (a) Randomly select 20 women for Group 1, which will see the "childcare" version of Company B's brochure, and assign the other 20 women to Group 2 (the "no childcare" group). Allow all women to examine the appropriate brochures, and observe which company they choose. Compare the number from Group 1 who choose Company B with the corresponding number from Group 2.

(b) Numbering from 01 to 40, Group 1 is 05-Cansico, 32-Roberts, 19-Hwang, 04-Brown, 25-Lippman, 29-Ng, 20-Iselin, 16-Gupta, 37-Turing, 39-Williams, 31-Rivera, 18-Howard, 07-Cortez, 13-Garcia, 33-Rosen, 02-Adamson, 36-Travers, 23-Kim, 27-McNeill, and 35-Thompson.

5.40 If this year is considerably different in some way from last year, we cannot compare electricity consumption over the two years. For example, if this summer is warmer, the customers may run their air conditioners more often. The possible differences between the two years would confound the effects of the treatments.

5.41 The second design is an experiment—a treatment is imposed on the subjects. The first is a study; it may be confounded by the types of men in each group. In spite of the researcher's attempt to match "similar" men from each group, those in the first group (who exercise) could be somehow different from men in the non-exercising group.

5.42 "Significantly higher" means that these returns are higher than would be expected due to random chance. "Not significantly different from zero" means that, although the sample may not be exactly zero, the difference from zero is within the range that can occur simply due to chance.

5.43 Because the experimenter knew which subjects had learned the meditation techniques, he (or she) may have had some expectations about the outcome of the experiment: if the experimenter believed that meditation was beneficial, he may subconsciously rate that group as being less anxious.

5.44 (a) If only the new drug is administered, and the subjects are then interviewed, their responses will not be useful, because there will be nothing to compare them to: How much "pain relief" does one expect to experience?

(b) Randomly assign 20 patients to each of three groups: Group 1, the placebo group; Group 2, the aspirin group; and Group 3, which will receive the new medication. After treating the patients, ask them how much pain relief they feel, and then compare the average pain relief experienced by each group.

(c) The subjects should certainly not know what drug they are getting—a patient told that she is receiving a placebo, for example, will probably not expect any pain relief.

(d) Yes—presumably, the researchers would like to conclude that the new medication is better than aspirin. If it is not double-blind, the interviewers may subtly influence the subjects into giving responses that support that conclusion.

5.45 (a) Ordered by increasing weight, the five blocks are (1) Williams-22, Deng-24, Hernandez-25, and Moses-25; (2) Santiago-27, Kendall-28, Mann-28, and Smith-29; (3) Brunk-30, Obrach-30, Rodriguez-30, and Loren-32; (4) Jackson-33, Stall-33, Brown-34, and Cruz-34; (5) Birnbaum-35, Tran-35, Nevesky-39, and Wilansky-42.

(b) The exact randomization will vary with the starting line in Table B. Different methods are possible; perhaps the simplest is to number from 1 to 4 within each block, then assign the members of block 1 to a weight-loss treatment, then assign block 2, etc. For example, starting on line 133, we assign 4-Moses to treatment A, 1-Williams to B, and 3-Hernandez to C (so that 2-Deng gets treatment D), then carry on for block 2, etc. (either continuing on the same line, or starting over somewhere else).

5.46 (a) Assume that the 6 circular areas are given in advance. Number them in any order. Use Table B to select 3 for the treatment. We used line 104. The first 4 digits are: 5 2 7 1. We cannot use the 7 because it is more than 6. Therefore, we would treat areas 5, 2 and 1.

(b) If the pairs are not given in advance, divide the 6 areas into 3 pairs so that the elements of each pair are close to each other and therefore of similar fertility. For each pair, we randomly pick one of the two to receive the treatment. Label the two areas in each pair A and B. If the random number from Table B is even, then apply the treatment to area A. Otherwise, apply the treatment to Area B. Alternatively, we could go along the table looking for either a 0 or a 1, ignoring the other digits. If we find a 0 before a 1, then treat area A. Otherwise, treat B.

5.47 (a) Assign 10 subjects to Group 1 (the 70° group) and the other 10 to Group 2 (which will perform the task in the 90° condition). Record the number of correct insertions in each group.

(b) All subjects will perform the task twice—once in each temperature condition. Randomly choose which temperature each subject works in first, either by flipping a coin, or by placing 10 subjects in Group 1 (70°, then 90°) and the other 10 in Group 2.

5.48 The randomization will vary with the starting line in Table B. *Completely randomized design:* Randomly assign 10 students to Group 1 (which has the trend-highlighting software) and the other 10 to Group 2 (which does not). Compare the performance of Group 1 with that of Group 2. *Matched pairs design:* Each student does the activity twice, once with the software and once without. Randomly decide (for each student) whether they have the software the first or second time. Compare performance with the software and without it. (This randomization can be done by flipping a coin 20 times, or by picking 20 digits from Table B, and using the software first if the digit is even, etc.) *Alternate matched pairs design:* Again, all students do the activity twice. Randomly assign 10 students to Group 1 and 10 to Group 2. Group 1 uses the software the first time; Group 2 uses the software the second time.

5.49 (a) "Randomized" means that patients were randomly assigned either St. John's wort or the placebo. "Placebo controlled" means that we will compare the results for the group using St. John's wort to the group that received the placebo.

(b)

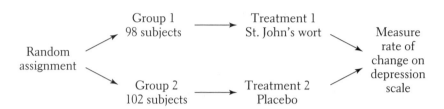

5.50 (a) The subjects are the 210 children. (b) The factor is the "choice set"; there are three levels (2 milk/2 fruit drink, 4 milk/2 fruit drink, and 2 milk/4 fruit drink). (c) The response variable is the choice made by each child.

5.51 (a) Outline:

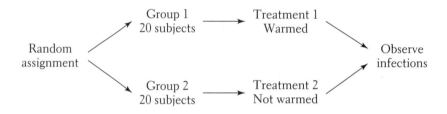

(b) Number the subjects from 01 to 40. Divide digits into groups of 2. Omit groups that are over 40. First 20 pairs will be Group 1. The rest will be Group 2.

(c) Assign each subject a group, 01 to 40, by alphabetical order. Starting at line 121 in Table B, the first twenty different groups we see are: 29-Ng, 07-Cordoba, 34-Sugiwara, 22-Kaplan, 10-Devlin, 25-Lucero, 13-Garcia, 38-Ullmann, 15-Green, 05-Cansico, 09-Decker, 08-Curzakis, 27-McNeill, 23-Kim, 30-Quinones, 28-Morse, 18-Howard, 03-Afifi, 01-Abbott, 36-Travers. These subjects will be assigned to Treatment Group 1; the remaining subjects go into Group 2.

(d) We want the treatment groups to be as alike as possible. If the same operating team was used to operate on "warmed" and "unwarmed" patients, then the effect of the "warming" on the occurrence of infection might be confounded with the effect of the surgical team (e.g., how skillful the team was in performing the necessary preventive measures).

(e) Double-blindness. We would prefer a double-blind experiment here to ensure that the patients would not be treated differently with regard to preventing and monitoring infections due to prior knowledge of how they were treated.

5.52

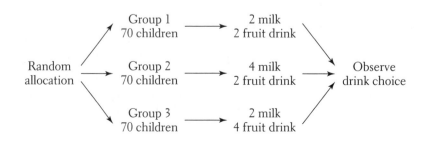

5.53 (a) Randomly assign 20 men to each of two groups. Record each subject's blood pressure, then apply the treatments: a calcium supplement for Group 1, and a placebo for Group 2. After sufficient time has passed, measure blood pressure again and observe any change.

(b) Number from 01 to 40 down the columns. Group 1 is 18-Howard, 20-Imrani, 26-Maldonado, 35-Tompkins, 39-Willis, 16-Guillen, 04-Bikalis, 21-James, 19-Hruska, 37-Tullock, 29-O'Brian, 07-Cranston, 34-Solomon, 22-Kaplan, 10-Durr, 25-Liang, 13-Fratianna, 38-Underwood, 15-Green, and 05-Chen.

(c) We prefer large treatment groups because differences in responses arising from such groups are more likely due to the treatments themselves, rather than chance variation.

5.54 Label the children from 001 to 210, then consider three digits at a time. The first five children in Group 1 are numbers 119, 033, 199, 192, and 148.

5.55 Responding to a placebo does not imply that the complaint was not "real"—38% of the placebo group in the gastric freezing experiment improved, and those patients really had ulcers. The placebo effect is a *psychological* response, but it may make an actual *physical* improvement in the patient's health.

5.56 (a) The explanatory variable is the vitamin(s) taken each day; the response variable is whether colon cancer develops.

(b) Diagram below; equal group sizes are convenient but not necessary.

(c) Using labels 001 through 864 (or 000 through 863), we choose 731, 253, 304, 470, and 296.

(d) "Double-blind" means that both the subjects and those who work with the subjects do not know who is getting what treatment. This prevents the expectations of those involved from affecting the way in which the subjects' conditions are diagnosed.

(e) The observed differences were no more than what might reasonably occur by chance even if there is no effect due to the treatments.

(f) Fruits and vegetables contain fiber; this could account for the benefits of those foods. Also, people who eat lots of fruit and vegetables may have healthier diets overall (e.g., less red meat).

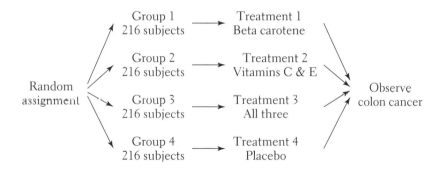

5.57 Three possible treatments are (1) fine, (2) jail time, and (3) attending counseling classes. The response variable would be the rate at which people in the three groups are rearrested.

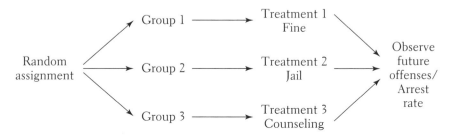

5.58 (a) Each subject takes both tests; the order in which the tests are taken is randomly chosen. (b) Take 22 digits from Table B. If the first digit is even, subject 1 takes the BI first; if it is odd, he or she takes the ARSMA first. (Or, administer the BI first if the first digit is 0–4, and the ARSMA first if it is 5–9.)

5.59 (a) Flip the coin twice. Let HH ⇔ failure, and let the other three outcomes, HT, TH, TT ⇔ success.

(b) Let 1,2,3 ⇔ success, and let 4 ⇔ failure. If 5 or 6 come up, ignore them and roll again.

(c) Peel off two consecutive digits from the table; let 01 through 75 ⇔ success, and let 76 through 99 and 00 ⇔ failure.

(d) Let diamond, spade, club ⇔ success, and let heart ⇔ failure.

5.60 Flip both nickels at the same time. Let HH ⇔ success (the occurrence of the phenomenon of interest) and HT, TH, TT ⇔ failure (the nonoccurrence of the phenomenon).

5.61 (a) Obtain an alphabetical list of the student body, and assign consecutive numbers to the students on the list. Use a random process (table or random digit generator) to select 10 students from this list.

(b) Let the two-digit groups 00 to 83 represent a "Yes" to the question of whether or not to abolish evening exams and the groups 84 to 99 represent a "No."

(c) Starting at line 129 in Table B ("Yes" in **boldface**) and moving across rows:

Repetition 1:	36, 75, 95, 89, 84, 68, 28, 82, 29, 13	# "Yes": 7
Repetition 2:	18, 63, 85, 43, 03, 00, 79, 50, 87, 27	# "Yes": 8
Repetition 3:	69, 05, 16, 48, 17, 87, 17, 40, 95, 17	# "Yes": 8
Repetition 4:	84, 53, 40, 64, 89, 87, 20, 19, 72, 45	# "Yes": 7
Repetition 5:	05, 00, 71, 66, 32, 81, 19, 41, 48, 73	# "Yes": 10

(Theoretically, we should achieve 10 "Yes" results approximately 10.7% of the time.)

5.62 (a) A single random digit simulates one shot, with 0 to 6 a hit and 7, 8, or 9 a miss. Then 5 consecutive digits simulate 5 independent shots.

(b) Let 0–6 ⇔ "hit" and 7, 8, 9 ⇔ "miss." Starting with line 125, the first four repetitions are:

$$9 \; \underline{6} \; 7 \; \underline{4} \; \underline{6} \qquad \underline{1} \; \underline{2} \; \underline{1} \; \underline{4} \; 9 \qquad \underline{3} \; 7 \; 8 \; \underline{2} \; \underline{3} \qquad 7 \; \underline{1} \; 8 \; \underline{6} \; 8$$
$$\qquad (3) \qquad\qquad\quad (4) \qquad\qquad\quad (3) \qquad\qquad\quad (2)$$

Each block of 5 digits in the table represents one repetition of the 5 attempted free throws. The underlined digits represent hits. We perform 46 more repetitions for a total of 50, and calculate the proportion of times the player makes 2 or fewer shots. Here are the number of hits for the 50 repetitions.

```
3  4  3  2  4      4  5  4  2  3      3  3  2  2  3      4  3  1  5  4
4  4  3  4  3      4  5  4  5  3      2  2  3  2  2      4  3  5  3  2
4  3  2  3  4      3  3  3  4  5
```

The frequency counts are

X	0	1	2	3	4	5
Freq.	0	1	10	18	15	6

The relative frequency of 2 or fewer hits in 5 attempts is 11/50 = .22.

Note: It will be shown in Chapter 8 that the theoretical probability of missing 3 or more shots (i.e., making 2 or fewer shots) is 0.1631, or about one time in six.

5.63 The choice of digits in these simulations may of course vary from that made here. In (a)–(c), a single digit simulates the response; for (d), two digits simulate the response of a single voter.

(a) Odd digits   — voter would vote Democratic
    Even digits   — voter would vote Republican

(b) 0, 1, 2, 3, 4, 5 — Democratic
    6, 7, 8, 9   — Republican

(c) 0, 1, 2, 3   — Democratic
    4, 5, 6, 7   — Republican
    8, 9   — Undecided

(d) 00, 01, ..., 52 — Democratic
    53, 54, ..., 99 — Republican

5.64 For the choices made in the solution to Exercise 5.63,

(a) D, R, R, R, R, R, R, D, R, D — 3 Democrats, 7 Republicans
(b) R, D, D, R, R, R, R, D, R, R — 3 Democrats, 7 Republicans
(c) R, U, R, D, R, U, U, U, D, R — 2 Democrats, 4 Republicans, 4 undecided
(d) R, R, R, D, D, D, D, D, D, R — 6 Democrats, 4 Republicans

5.65 Let 1 = girl and 0 = boy. The command `randInt(0, 1)` produces a 0 or 1 with equal likelihood. Continue to press ENTER. In 50 repetitions, we got a girl 47 times, and all 4 boys three times. Our simulation produced a girl 94% of the time, vs. a theoretical probability of 0.938.

5.66 (a) Let the digits 0, 1, 2, 3, 4, 5 ⇔ the American League team winning a Series game and 6, 7, 8, 9 ⇔ the National League team winning. We choose single digits until one team has won four games, with a minimum of four digits and a maximum of seven digits being chosen. On the TI-83, you can use the command `randInt (0, 9, 1)` repeatedly to generate the digits. Here are several sample simulations:

| 0, 3, 9, 2, 7, 9, 2 | AL, AL, NL, AL, NL, NL, AL | # games = 7 |
| 3, 0, 9, 1, 0 | AL, AL, NL, AL, AL | # games = 5 |

The long-term average of many simulations will give the approximate number of games one would expect the Series to last.

(b) Other factors might include: the starting pitchers, the weather conditions, the injury status of key players.

5.67 Let 01 to 15 ⇔ breaking a racquet, and let 16 to 99 and 00 ⇔ not breaking a racquet. Starting with line 141 in the random digit table, we peel two digits off at a time and record the results: 96 76 73 59 64 23 82 29 60 12. In the first repetition, Brian played 10 matches until he broke a racquet. Addition repetitions produced these results: 3, 11, 6, 37, 5, 3, 4, 11, 1. The average for these 10 repetitions is 9.1. We will learn in Chapter 8 that the expected number of matches until a break is about 6.67. More repetitions should improve our estimate.

5.68 (a) Read two random digits at a time from Table B. Let 01 to 13 represent a Heart, let 14 to 52 represent another suit, and ignore the other two-digit numbers. (b) You should beat Slim about 44% of the time.

5.69 On the TI-83, we started a counter (C), and then executed the command shown, pressing the ENTER key 30 times for 30 repetitions.

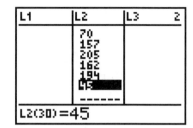

For five sets of 30 repetitions, we observed 5, 3, 3, 8, and 4 numbers that were multiples of 5. The mean number of multiples of 5 in 30 repetitions was 3.6, so 3.6/30 = 12% is our estimate for the proportion of times a person wins the game.

5.70 The command randInt (1, 365, 23) → L_1 : SortA (L_1) randomly selects 23 birthdays and assigns them to L_1. Then it sorts the day in increasing order. Scroll through the list to see duplicate birthdays. Repeat many times. For a large number of repetitions, there should be duplicate birthdays about half the time. To simulate 41 people, change 23 to 41 in the command and repeat many times. We assume that there are 365 days for birthdays, and that all days are equally likely to be a birthday.

5.71 (a) Let 000 to 999 ⇔ a bats, 000 to 319 ⇔ hits, and 320 to 999 ⇔ no hits.

(b) We entered 1 → c ENTER to set a counter. Then enter randInt (0, 999, 20) → L_1 : sum (L_1 ≥ 0 and L_1 ≤ 319) → L_2 (C) : C + 1 > C and press ENTER repeatedly. The count (number of the repetition) is displayed on the screen to help you see when to stop. The results for the 20 repetitions are stored in list L_2. We obtained the following frequencies:

Number of hits in 20 at bats	4	5	6	7	8	9
Frequency	3	5	4	3	2	3

(c) The mean number of hits in 20 at bats was $\bar{x}$ = 6.25. And 6.25/20 = .3125, compared with the player's batting average of .320. Notice that even though there was considerable variability in the 20 repetitions, ranging from a low of 3 hits to a high of 9 hits, the results of our simulation were very close to the player's batting average.

5.72 (a) One digit simulates system A's response: 0 to 8 shut down the reactor, and 9 fails to shut it down.

(b) One digit simulates system B's response: 0 to 7 shut down the reactor, and 8 or 9 fail.

(c) A pair of consecutive digits simulates the response of both systems, the first giving A's response as in (a), and the second B's response as in (b). If a single digit were used to simulate both systems, the reactions of A and B would be dependent—for example, if A fails, then B must also fail.

(d) The true probability that the reactor will shut down is 1− (0.2)(0.1) = 0.98.

5.73 This simulation is fun for students, but the record-keeping can be challenging! Here is one method. First number the (real or imaginary) participants 1–25. Write the numbers 1–25 on the board so that you can strike through them as they hear the rumor. We used randInt (1, 25) to randomly select a person to begin spreading the rumor, and then pressed ENTER repeatedly to randomly select additional people to hear the rumor. We made a table to record the round (time increment), those who knew the rumor and were spreading it, those randomly selected to hear the rumor, and those who stopped spreading it because the person randomly selected to hear it had already heard it. Here is the beginning of our simulation, to illustrate our scheme:

Time incr	Knows		Tells	Stopped
1	16	→	2	
2	2	→	25	
	16	→	3	
3	2		19	
	3		6	
	16		15	
	25		1	
4	1		21	
	2		5	
	3		23	
	6		13	
	15		25	15
	16		9	
	19		16	19
	25		15	25
5				

Eventually we crossed off all but 7, 12, 14, and 24, so 4 out of 25 or 4/25 = 16% never heard the rumor. It can be shown that with a sufficiently large population, approximately 20% of the population will not hear the rumor.

5.74 (a) The population is Ontario residents; the sample is the 61,239 people interviewed. (b) The sample size is very large, so if there were large numbers of both sexes in the sample—this is a safe assumption since we are told this is a "random sample"—these two numbers should be fairly accurate reflections of the values for the whole population.

5.75 (a) Explanatory variable: treatment method; response: survival times. (b) No treatment is actively imposed; the women (or their doctors) chose which treatment to use. (c) Doctors may make the decision of which treatment to recommend based in part on how advanced the case is. Some might be more likely to recommend the older treatment for advanced cases, in which case the chance of recovery is lower. Other doctors might view the older treatment as not being worth the effort, and recommend the newer method as a way of providing some hope for recovery while minimizing the trauma and expense of major surgery.

5.76 (a) Sample survey. (b) Experiment. The treatment would be classroom or online course. (c) Observational study.

5.77 Divide the players into two groups—one to receive oxygen and the other without oxygen during the rest period. Match the players so each player receiving oxygen has a corresponding player of similar speed who does not receive oxygen.

5.78 A stratified random sample would be useful here; one could select 50 faculty members from each level. Alternatively, select 25 (or 50) institutions of each size, then choose 2 (or 1) faculty members at each institution.
   If a large proportion of faculty in your state work at a particular class of institution, it may be useful to stratify unevenly. If, for example, about 50% teach at Class I institutions, you may want half your sample to come from Class I institutions.

5.79 (a) The chicks are the experimental units; weight gain is the response variable.

(b) There are two factors: corn variety (2 levels) and percent of protein (3 levels). This makes 6 treatments, so 60 chicks are required.

| | | Factor B: Protein level | | |
		12%	16%	20%
Factor A:	opaque-2	1	2	3
Corn variety	floury-2	4	5	6

(c)

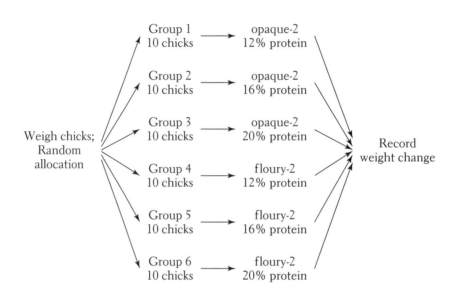

5.80 (a) This is a randomized block design. The blocks here are "runners" and "nonrunners."

(b)

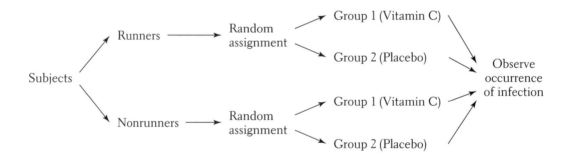

(c) A difference in rate of infection may have been due to the effects of the treatments, or it may simply have been due to random chance. Saying that the placebo rate of 68% is "significantly more" than the Vitamin C rate of 33% means that the observed difference is too large to have occurred by chance alone. In other words, Vitamin C appears to have played a role in lowering the infection rate of runners.

5.81 The factors are whether or not the letter has a ZIP code (2 levels: yes or no), and the time of day the letter is mailed. The number of levels for the second factor may vary.

To deal with lurking variables, all letters should be the same size and should be sent to the same city, and the day on which a letter is sent should be randomly selected. Because most post offices have shorter hours on Saturdays, one may wish to give that day some sort of "special treatment" (it might even be a good idea to have the day of the week be a *third* factor in this experiment).

5.82 Each subject should taste both kinds of cheeseburger, in a randomly selected order, and then be asked about their preference. Both burgers should have the same "fixings" (ketchup, mustard, etc.). Since some subjects might be able to identify the cheeseburgers by appearance, one might need to take additional steps (such as blindfolding, or serving only the center part of the burger) in order to make this a truly "blind" experiment.

5.83 (a)

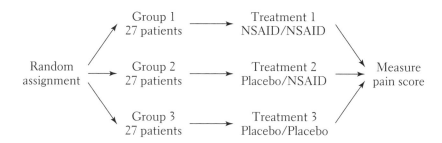

The two extra patients can be randomly assigned to two of the three groups.

(b) No one involved in administering treatments or assessing their effectiveness knew which subjects were in which group.

(c) The pain scores in Group A were so much lower than the scores in Groups B and C that they would not often happen by chance if NSAIDs were not effective. We can conclude that NSAIDs provide real pain relief.

5.84 (a) A single run: spin the 1–10 spinner twice; see if the larger of the two numbers is larger than 5. The player wins if either number is 6, 7, 8, 9, or 10.

(b) If using the random digit table, let 0 represent 10, and let the digits 1–9 represent themselves.

(c) `randInt (1, 10, 2)`.

(d) In our simulation of 20 repetitions, we observed 13 wins for a 65% win rate. Using the methods of the next chapter, it can be shown that there is a 75% probability of winning this game.

5.85 (a) Let 01 to 05 represent demand for 0 cheesecakes
Let 06 to 20 represent demand for 1 cheesecake
Let 21 to 45 represent demand for 2 cheesecakes
Let 46 to 70 represent demand for 3 cheesecakes
Let 71 to 90 represent demand for 4 cheesecakes
Let 91 to 99 and 00 represent demand for 5 cheesecakes

(b) Our results suggest that the baker should make 2 cheesecakes each day to maximize his profits.

5.86 (a) Since Carla makes 80% of her free throws, let a single digit represent a free throw, and let $0–7 \Leftrightarrow$ "hit" and $8, 9 \Leftrightarrow$ "miss."

(b) We instructed the calculator to simulate a free throw, and store the result in L_1. Then we instructed the calculator to see if the attempt was a hit (1) or a miss (0), and record that fact in L_2. Continue to press ENTER until there are 20 simulated free throws.

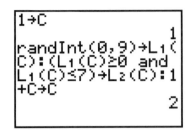

Scroll through L_2 and determine the longest string of 1's (consecutive baskets). In our first set of 20 repetitions, we observed 8 consecutive baskets. Additional sets of 20 repetitions produced: 11, 9, 9, and 8. The average longest run was 9 consecutive baskets in 20 attempts. The five number summary: Min = 8, Q_1 = 8, Med = 9, Q_3 = 10, and Max = 11. There was surprisingly little variation in our five groups of 20 repetitions.

5.87 (a) A single digit simulates one try, with 0 or 1 a pass and 2 to 9 a failure. Three independent tries are simulated by three successive random digits.

(b) With the convention of (a), 50 tries beginning in line 120 gives 25 successes, so the probability of success is estimated as 25/50 = 1/2. [In doing the simulation, remember that you can end a repetition after 1 or 2 tries if the student passes, so that some repetitions do not use three digits. Though this is a proper simulation of the student's behavior, the probability of at least one pass is the same if three digits are examined in every repetition. The true probability is $1 - (0.8)^3$ = 0.488, so this particular simulation was quite accurate.]

(c) No—learning usually occurs in taking an exam, so the probability of passing probably increases on each trial.

5.88 (a) The three tries are simulated by three consecutive random digits (stopping after a pass): 0 and 1 are a pass on the first try; 0, 1, and 2 a pass on the second try; and 0, 1, 2, and 3 a pass on the third try.

(b) The correct probability is $1 - (0.8)(0.7)(0.6) = 0.664$. Taking groups of three digits at a time (not quitting early after a pass) gives 36 passes, so the estimated probability is 0.72. If one does quit early after a pass, there are 32 passes, so the estimated probability is 0.64.

6

Probability: The Study of Randomness

6.1 Long trials of this experiment often approach 40% heads. One theory attributes this surprising result to a "bottle-cap effect" due to an unequal rim on the penny. We don't know. But a teaching assistant claims to have spent a profitable evening at a party betting on spinning coins after learning of the effect.

6.2 The theoretical probabilities are, in order: $1/16$, $4/16 = 1/4$, $6/16 = 3/8$, $4/16 = 1/4$, $1/16$.

6.3 (b) In our simulation, Shaq hit 52% of his shots. (c) The longest sequence of misses in our run was 6 and the longest sequence of hits was 9. Of course, results will vary.

6.4 (a) 0. (b) 1. (c) 0.01. (d) 0.6 (or 0.99, but "more often than not" is a rather weak description of an event with probability 0.99!)

6.5 There are 21 0s among the first 200 digits; the proportion is $\frac{21}{200} = 0.105$.

6.6 (a) We expect probability $1/2$ (for the first flip, or for *any* flip of the coin). (b) The theoretical probability that the first head appears on an odd-numbered toss of a fair coin is $\frac{1}{2} + \left(\frac{1}{2}\right)^3 + \left(\frac{1}{2}\right)^5 + \ldots = \frac{2}{3}$. Most answers should be between about 0.47 and 0.87.

6.7 Obviously, results will vary with the type of thumbtack used. If you try this experiment, note that although it is commonly done when flipping coins, we do not recommend throwing the tack in the air, catching it, and slapping it down on the back of your other hand.

6.8 In the long run, of a large number of hands of five cards, about 2% (one out of 50) will contain a three of a kind. (Note: This probability is actually $\frac{88}{4165} \doteq 0.02113$.)

6.9 The study looked at regular season games, which included games against poorer teams, and it is reasonable to believe that the 63% figure is inflated because of these weaker opponents. In the World Series, the two teams will (presumably) be nearly the best, and home game wins will not be so easy.

6.10 (a) With $n = 20$, nearly all answers will be 0.40 or greater. With $n = 80$, nearly all answers will be between 0.58 and 0.88. With $n = 320$, nearly all answers will be between 0.66 and 0.80.

6.11 (a) S = {germinates, fails to grow}. (b) If measured in weeks, for example, S = {0, 1, 2, ...}. (c) S = {A, B, C, D, F}. (d) Using Y for "yes (shot made)" and N for "no (shot missed)," S = {YYYY, NNNN, YYYN, NNNY, YYNY, NNYN, YNYY, NYNN, NYYY, YNNN, YYNN, NNYY, YNYN, NYNY, YNNY, NYYN}. (There are 16 items in the sample space.) (e) S = {0, 1, 2, 3, 4}.

6.12 (a) S = {all numbers between 0 and 24}. (b) S = {0, 1, 2, ... .11000}. (c) S = {0, 1, 2, ..., 12}. (d) S = {all numbers greater than or equal to 0}, or S = {0, 0.01, 0.02, 0.03, ...}. (e) S = {all positive and negative numbers}. Note that the rats can lose weight.

6.13 S = {all numbers between _____ and _____}. The numbers in the blanks may vary. Table 1.10 has values from 86 to 195 cal; the range of values in S should include *at least* those numbers. Some students may play it safe and say "all numbers greater than 0."

6.14 If two coins are tossed, then by the multiplication principle, there are $(2)(2) = 4$ possible outcomes. The outcomes are illustrated in the following tree diagram:

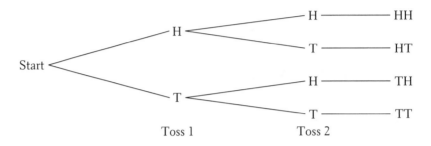

The sample space is {HH, HT, TH, TT}. (b) If three coins are tossed, then there are $(2)(2)(2) = 8$ possible outcomes. The outcomes are illustrated in the following tree diagram:

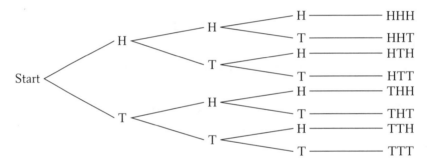

The sample space is {HHH, HHT, HTH, HTT, THH, THT, TTH, TTT}. (c) If four coins are tossed, then there are $(2)(2)(2)(2) = 16$ possible outcomes, each of which consists of a string of four letters that may be H's or T's. The sample space is {HHHH, HHHT, HHTH, HTHH, THHH, HHTT, HTHT, HTTH, THTH, TTHH, THHT, HTTT, THTT, TTHT, TTTH, TTTT}.

6.15 (a) $10 \times 10 \times 10 \times 10 = 10^4 = 10,000$. (b) $10 \times 9 \times 8 \times 7 = 5,040$. (c) There are 10,000 four-digit tags, 1,000 three-digit tags, 100 two-digit tags, and 10 one-digit tags, for a total of 11,110 license tags.

6.16 (a) An outcome of this experiment consists of a string of 3 digits, each of which can be 1, 2, or 3. By the multiplication principle, the number of possible outcomes is $(3)(3)(3) = 27$. (b) The sample space is {111, 112, 113, 121, 122, 123, 131, 132, 133, 211, 212, 213, 221, 222, 223, 231, 232, 233, 311, 312, 313, 321, 322, 323, 331, 332, 333}.

6.17 (a)

Number of ways	Sum	Outcomes
1	2	(1, 1)
2	3	(1, 2) (2, 1)
3	4	(1, 3) (2, 2) (3, 1)
4	5	(1, 4) (2, 3) (3, 2) (4, 1)
5	6	(1, 5) (2, 4) (3, 3) (4, 2) (5, 1)
6	7	(1, 6) (2, 5) (3, 4) (4, 3) (5, 2) (6, 1)
5	8	(2, 6) (3, 5) (4, 4) (5, 3) (6, 2)
4	9	(3, 6) (4, 5) (5, 4) (6, 3)
3	10	(4, 6) (5, 5) (6, 4)
2	11	(5, 6) (6, 5)
1	12	(6, 6)

(b) 18.

(c) There are 4 ways to get a sum of 5 and 5 ways to get a sum of 8.

(d) Answers will vary but might include:

- The "number of ways" increases until "sum = 7" and then decreases.
- The "number of ways" is symmetrical about "sum = 7."
- Odd sums occur an even number of ways and even sums occur an odd number of ways.

6.18 (a) 26. (b) 13. (c) 1. (d) 16. (e) 3.

6.19 (a) The given probabilities have sum 0.96, so P(type AB) = 0.04. (b) P(type O or B) = 0.49 + 0.20 = 0.69.

6.20 (a) The sum of the given probabilities is 0.9, so P(blue) = 0.1. (b) The sum of the given probabilities is 0.7, so P(blue) = 0.3. (c) P(plain M&M is red, yellow, or orange) = 0.2 + 0.2 + 0.1 = 0.5. P(peanut M&M is red, yellow, or orange) = 0.1 + 0.2 + 0.1 = 0.4.

6.21 P(either CV disease or cancer) = 0.45 + 0.22 = 0.67; P(other cause) = $1 - 0.67 = 0.33$.

6.22 (a) Since the three probabilities must add to 1 (assuming that there were no "no opinion" responses), this probability must be $1 - (0.12 + 0.61) = 0.27$. (b) 0.12 + 0.61 = 0.73.

6.23 (a) The sum is 1, as we expect since all possible outcomes are listed. (b) $1 - 0.41 = 0.59$. (c) 0.41 + 0.23 = 0.64. (d) (0.41)(0.41) = 0.1681.

6.24 There are 19 outcomes where at least one digit occurs in the correct position: 111, 112, 113, 121, 122, 123, 131, 132, 133, 213, 221, 222, 223, 233, 313, 321, 322, 323, 333. The theoretical probability of at least one digit occurring in the correct position is therefore 19/27 = .7037.

6.25 (a) Let x = number of spots. Then $P(x = 1) = P(x = 2) = P(x = 3) = P(x = 4) = 0.25$. Since all 4 faces have the same shape and the same area, it is reasonable to assume that the probability of a face being down is the same as for any other face. Since the sum of the probabilities must be one, the probability of each should be 0.25.

(b) Outcomes (1,1) (1,2) (1,3) (1,4) (2,1) (2,2) (2,3) (2,4) (3,1) (3,2) (3,3) (3,4) (4,1) (4,2) (4,3) (4,4) The probability of any pair is 1/16 = 0.0625.
P(Sum = 5) = P(1,4) + P(2,3) + P(3,2) + P(4,1) = (0.0625) (4) = 0.25.

6.26 (a) $P(D) = P(1, 2, \text{ or } 3) = .301 + .176 + .125 = .602$.

(b) $P(B \cup D) = P(B) + P(D) = .602 + .222 = .824$.

(c) $P(D^c) = 1 - P(D) = 1 - .602 = .398$.

(d) $P(C \cap D) = P(1 \text{ or } 3) = .301 + .125 = .426$.

(e) $P(B \cap C) = P(7 \text{ or } 9) = .058 + .046 = .104$.

6.27 Fight one big battle: His probability of winning is 0.6, compared to $0.8^3 = 0.512$. (Or he could choose to try for a negotiated peace.)

6.28 $(1 - 0.05)^{12} = (0.95)^{12} = 0.5404$.

6.29 No: It is unlikely that these events are independent. In particular, it is reasonable to expect that college graduates are less likely to be laborers or operators.

6.30 (a) $P(A) = \frac{38,225}{166,438} = 0.230$ since there are 38,225 (thousand) people who have completed 4+ years of college out of 166,438 (thousand). (b) $P(B) = \frac{52,022}{166,438} = 0.313$. (c) $P(A \text{ and } B) = \frac{8,005}{166,438} = 0.048$; A and B are not independent since $P(A \text{ and } B) \neq P(A)P(B)$.

6.31 An individual light remains lit for 3 years with probability $1 - 0.02$; the whole string remains lit with probability $(1 - 0.02)^{20} = (0.98)^{20} \doteq 0.6676$.

6.32 P(neither test is positive) = $(1 - 0.9)(1 - 0.8) = (0.1)(0.2) = 0.02$.

6.33 (a) P(one call does not reach a person) = 0.8. Thus, P(none of the 5 calls reaches a person) = $(0.8)^5 = 0.32768$.

(b) P(one call to NYC does not reach a person) = 0.92. Thus, P(none of the 5 calls to NYC reach a person) = $(0.92)^5$ = 0.6591.

6.34 Model 1: Legitimate. Model 2: Legitimate. Model 3: Probabilities have sum $\frac{6}{7}$. Model 4: Probabilities cannot be negative.

6.35 (a) Legitimate. (b) Not legitimate, because probabilities sum to more than 1. (c) Not legitimate, because probabilities sum to less than 1.

6.36 (a) Sum of given probabilities = .705, so P(car has some other color) = 1 − .705 = .295.
(b) P(silver or white) = P(silver) + P(white) = .176 + .172 = .348.
(c) Assuming that the vehicle choices are independent, P(both silver or white) = $(.348)^2$ = .121.

6.37 (a) The sum of all 8 probabilities equals 1 and all probabilities satisfy $0 \le p \le 1$.
(b) $P(A)$ = 0.000 + 0.003 + 0.060 + 0.062 = 0.125.
(c) The chosen person is not white.
$$P(B^c) = 1 - P(B) = 1 - (0.060 + 0.691) = 1 - 0.751 = 0.249.$$
(d) $P(A^c \cap B)$ = 0.691.

6.38 A, B are not independent because $P(A \text{ and } B)$ = 0.06, but $P(A) \times P(B)$ = (.125)(.751) = 0.093875. For the events to be independent, these two probabilities must be equal.

6.39 (a) P(undergraduate and score $\ge$ 600) = (0.40)(0.50) = 0.20.
P(graduate and score $\ge$ 600) = (0.60)(0.70) = 0.42.
(b) P(score $\ge$ 600) = P(UG and score $\ge$ 600) + P(G and score $\ge$ 600) = 0.20 + 0.42 = 0.62.

6.40 (a) $(0.65)^3$ = 0.2746 (under the random walk theory). (b) 0.35 (since performance in separate years is independent). (c) $(0.65)^2 + (0.35)^2$ = 0.545.

6.41 (a) P(under 65) = 0.321 + 0.124 = 0.445. P(65 or older) = 1 − 0.445 = 0.555. (b) P(tests done) = 0.321 + 0.365 = 0.686. P(tests not done) = 1 − 0.686 = 0.314. (c) $P(A \text{ and } B)$ = 0.365; $P(A)P(B)$ = (0.555)(0.686) = 0.3807. A and B are not independent; tests were done less frequently on older patients than if these events were independent.

6.42 (a) 1/38. (b) Since 18 slots are red, the probability of a red is P(red) = $\frac{18}{38}$ = 0.474. (c) There are 12 winning slots, so P(win a column bet) = $\frac{12}{38}$ = 0.316.

6.43 Look at the first five rolls in each sequence. All have one G and four R's, so those probabilities are the same. In the first sequence, you win regardless of the sixth roll; for the second, you win if the sixth roll is G, and for the third sequence, you win if it is R. The respective probabilities are $\left(\frac{2}{6}\right)^4 \cdot \left(\frac{4}{6}\right)$
= $\frac{2}{243}$ = 0.00823, $\left(\frac{2}{6}\right)^4 \cdot \left(\frac{4}{6}\right)^2$ = $\frac{4}{729}$ = 0.00549, and $\left(\frac{2}{6}\right)^5 \cdot \left(\frac{4}{6}\right)$ = $\frac{2}{729}$ = 0.00274.

6.44 P(first child is albino) = $\frac{1}{2} \times \frac{1}{2} = \frac{1}{4}$. P(both of two children are albino) = $\frac{1}{4} \times \frac{1}{4} = \frac{1}{16}$. P(neither is albino) = $\left(1 - \frac{1}{4}\right)^2 = \frac{9}{16}$.

6.45 (a) If A, B are independent, then $P(A \text{ and } B) = P(A) \cdot P(B)$. Since A and B are nonempty, then we have $P(A) > 0$, $P(B) > 0$ and $P(A) \cdot P(B) > 0$. Therefore, $P(A \text{ and } B) > 0$. So A and B cannot be empty.
(b) If A and B are disjoint, then $P(A \text{ and } B) = 0$. But this cannot be true if A and B are independent by part (a). So A and B cannot be independent.
(c) Example: A bag contains 3 red balls and 2 green balls. A ball is drawn from the bag, its color is noted, and the ball is set aside. Then a second ball is drawn and its color is noted. Let event A be the event that the first ball is red. Let event B be the event that the second ball is red. Events A and B are not disjoint because both balls can be red. However, events A and B are not independent because whether the first ball is red or not, alters the probability of the second ball being red.

6.46 $P(A \text{ or } B) = P(A) + P(B) - P(A \text{ and } B) = 0.125 + 0.237 - 0.077 = 0.285.$

6.47

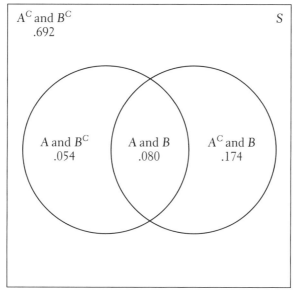

(a) {A and B} represents both prosperous and educated. $P(A \text{ and } B) = 0.080$. (b) {A and B^C} represents prosperous but not educated. $P(A \text{ and } B^C) = P(A) - P(A \text{ and } B) = .134 - .080 = .054.$ (c) {A^C and B} represents not prosperous but educated. $P(A^C \text{ and } B) = P(A) - P(A \text{ and } B) = .254 - .080 = .174.$ (d) {A^C and B^C} represents neither prosperous nor educated. $P(A^C \text{ and } B^C) = 1 - (.054 + .080 + .174) = 1 - .308 = .692.$

6.48 $P(A \text{ or } B) = P(A) + P(B) - P(A \text{ and } B) = 0.6 + 0.4 - 0.2 = 0.8.$

6.49 $P(A) \cdot P(B) = (0.6)(0.5) = 0.30$
Since this equals the stated probability for $P(A \text{ and } B)$, events A and B are independent.

6.50 (a) This event is {A and B}; $P(A \text{ and } B) = 0.2$ (given). (b) This is {A and B^c}; $P(A \text{ and } B^c) = P(A) - P(A \text{ and } B) = 0.4.$ (c) This is {A^c and B}; $P(A^c \text{ and } B) = P(B) - P(A \text{ and } B) = 0.2.$ (d) This is {A^c and B^c}; $P(A^c \text{ and } B^c) = 0.2$ (so that the probabilities add to 1).

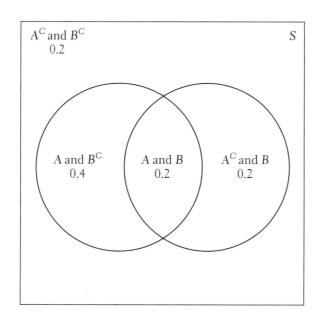

6.51 In constructing the Venn diagram, start with the numbers given for "only tea" and "all three," then determine other values. For example, P(coffee and cola, but not tea) = P(coffee and cola) − P(all three). (a) 15% drink only cola. (b) 20% drink none of these.

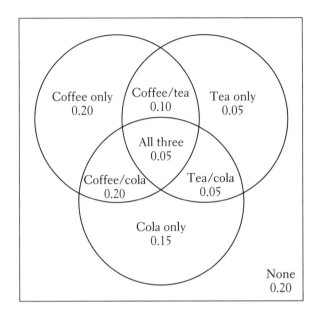

6.52 (a) Below. (b) P(country but not Gospel) = $P(C) − P(C$ and $G) = 0.4 − 0.1 = 0.3$. (c) P(neither) = $1 − P(C$ or $G) = 1 − (0.4 + 0.3 − 0.1) = 0.4$.

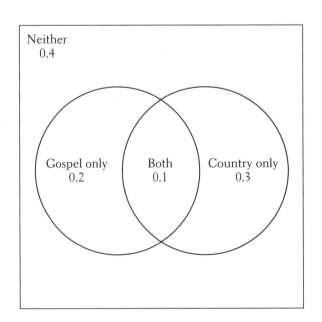

6.53 (a) On the facing page. (B) P(neither admits Ramon) = $1 − P(P$ or $S) = 1 − (0.4 + 0.5 − 0.2) = 0.3$. (c) $P(S$ and not $P) = P(S) − P(P$ and $S) = 0.3$.

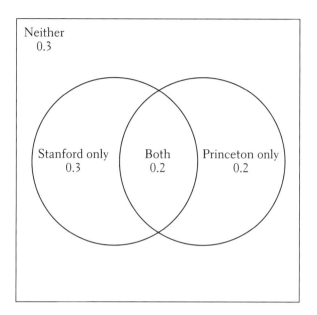

6.54 (a) 18669/103870 = .18.

(b) 8270/18669 = .443.

(c) 8270/103870 = .08.

6.55 (a) 7842/59920 = 0.13087.

(b) Married, age 18 to 29.

(c) 0.13087 is the proportion of women who are age 18 to 29 among those women who are married.

6.56 $P(A \text{ and } B) = P(A) P(B \mid A) = (0.46)(0.32) = 0.1472$.

6.57 If $F = \{\text{dollar falls}\}$ and $R = \{\text{renegotiation demanded}\}$, then $P(F \text{ and } R) = P(F) P(R \mid F) = (0.4)(0.8) = 0.32$.

6.58 (a) & (b) These probabilities are below. (c) The product of these conditional probabilities gives the probability of a flush in spades by the extended multiplication rule: We must draw a spade, and then another, and then a third, a fourth, and a fifth. The product of these probabilities is about 0.0004952. (d) Since there are four possible suits in which to have a flush, the probability of a flush is four times that found in (c), or about 0.001981.

$$P(\text{1st card } \spadesuit) = \tfrac{13}{52} = \tfrac{1}{4} = 0.25$$

$$P(\text{2nd card } \spadesuit \mid 1\spadesuit \text{ picked}) = \tfrac{12}{51} = \tfrac{4}{17} \doteq 0.2353$$

$$P(\text{3rd card } \spadesuit \mid 2\spadesuit\text{s picked}) = \tfrac{11}{50} \qquad \doteq 0.22$$

$$P(\text{4th card } \spadesuit \mid 3\spadesuit\text{s picked}) = \tfrac{10}{49} \qquad \doteq 0.2041$$

$$P(\text{5th card } \spadesuit \mid 4\spadesuit\text{s picked}) = \tfrac{9}{48} = \tfrac{3}{16} = 0.1875$$

6.59 First, concentrate on (say) spades. The probability that the first card dealt is one of those five cards (A$\spadesuit$, K$\spadesuit$, Q$\spadesuit$, J$\spadesuit$, or 10$\spadesuit$) is 5/52. The conditional probability that the second is one of those cards, given that the first was, is 4/51. Continuing like this, we get 3/50, 2/49, and finally 1/48; the product of these five probabilities gives $P(\text{royal flush in spades}) \doteq 0.00000038477$. Multiplying by four gives $P(\text{royal flush}) \doteq 0.000001539$.

6.60 (a) $P(\text{income} \geq \$50,000) = 0.20 + 0.05 = 0.25$.

(b) $P(\text{income} \geq \$100,000 \mid \text{income} \geq \$50,000) = 0.05/0.25 = 0.2$.

6.61 Let $G = \{\text{student likes Gospel}\}$ and $C = \{\text{student likes country}\}$. See the Venn diagram in the solution to Exercise 6.52. (a) $P(G \mid C) = P(G \text{ and } C)/P(C) = 0.1/0.4 = 0.25$. (b) $P(G \mid \text{not } C) = P(G \text{ and not } C)/P(\text{not } C) = 0.2/0.6 = \frac{1}{3} \doteq 0.33$.

6.62 (a) $P(\text{at least } \$100,000) = 9{,}534{,}653/127{,}075{,}145 = .075$; $P(\text{at least } \$1 \text{ million}) = 205{,}124/127{,}075{,}145 = .002$.

(b) $P(\text{at least } \$1 \text{ million} \mid \text{at least } \$100,000) = .002/.075 = .027$.

6.63 Let $I = $ the event that the operation results in infection, $F = $ the event that the operation fails. We seek $P(I^c \text{ and } F^c)$. We are given that $P(I) = .03$, $P(F) = .14$, and $P(I \text{ and } F) = .01$. Then $P(I^c \text{ and } F^c) = 1 - P(I \text{ or } F) = 1 - (.03 + .14 - .01) = .84$.

6.64 (a)

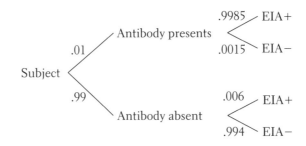

(b) $P(\text{test pos}) = P(\text{antibody and test pos}) + P(\text{no antibody and test pos.}) = (.01)(.9985) + (.99)(.006) = .016$.

(c) $P(\text{antibody} \mid \text{test pos}) = P(\text{antibody and test pos})/P(\text{test pos}) = (.01)(.9985)/.016 = .624$

6.65 (a) $P(\text{antibody} \mid \text{test pos}) = \frac{0.0009985}{0.0009985 + 0.005994} = 0.1428$.

(b) $P(\text{antibody} \mid \text{test pos}) = \frac{0.09985}{0.09985 + 0.0054} = 0.9487$.

(c) A positive result does not always indicate that the antibody is present. How common a factor is in the population can impact the test probabilities.

6.66 (a) $P(\text{chemistry}) = 110/445 = .247$. 24.7% of all laureates won prizes in chemistry.

(b) $P(\text{US}) = 198/445 = .445$. 44.5% of all laureates did research in the United States.

(c) $P(\text{US} \mid \text{phys-med}) = 82/144 = .569$. 56.9% of all physiology/medicine laureates did research in the United States.

(d) $P(\text{phys-med} \mid \text{US}) = 82/198 = .414$. 41.4% of all laureates from the United States won prizes in physiology/medicine.

6.67 (a) $\frac{856}{1{,}626} = 0.5264$. (b) $\frac{30}{74} = 0.4054$. (c) No: If they were independent, the answers to (a) and (b) would be the same.

6.68 (a) $P(\text{jack}) = 1/13$.

(b) $P(5 \text{ on second} \mid \text{jack on first}) = 1/12$.

(c) $P(\text{jack on first and 5 on second}) = P(\text{jack on first}) \times P(5 \text{ on second} \mid \text{jack on first}) = (1/13) \times (1/12) = 1/126$.

(d) $P(\text{both cards greater than 5}) = P(\text{first card greater than 5}) \times P(\text{second card greater than 5} \mid \text{first card greater than 5}) = (8/13) \times (7/12) = 56/126 = 4/9$.

6.69 (a) $\frac{770}{1626} \doteq 0.4736$. (b) $\frac{529}{770} = 0.6870$. (d) Using multiplication rule: P(male and bachelor's degree) = P(male)P(bachelor's degree | male) = $(0.4736)(0.6870) = 0.3254$. (Answers will vary with how much previous answers had been rounded.) Directly: $\frac{529}{1626} \doteq 0.3253$. [Note that the difference between these answers is inconsequential, since the numbers in the table are rounded.]

6.70 (a) $P(C) = 0.20$, $P(A) = 0.10$, $P(A \mid C) = 0.05$.
(b) $P(A \text{ and } C) = P(C) \times P(A \mid C) = (0.20)(0.05) = 0.01$.

6.71 Percent of "A" students involved in an accident is

$$P(C \mid A) = \frac{P(C \text{ and } A)}{P(A)} = \frac{0.01}{0.10} = 0.10.$$

6.72 If $F = \{$dollar falls$\}$ and $R = \{$renegotiation demanded$\}$, then $P(R) = P(F \text{ and } R) + P(F^c \text{ and } R) = 0.32 + P(F^c) \, P(R \mid F^c) = 0.32 + (0.6)(0.2) = 0.44$.

6.73 $P(\text{correct}) = P(\text{knows answer}) + P(\text{doesn't know, but guesses correctly}) = 0.75 + (0.25)(0.20) = 0.8$.

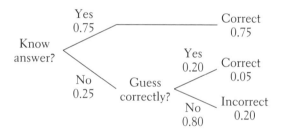

6.74 Tree diagram below. The black candidate expects to get 12% + 36% + 10% = 58% of the vote.

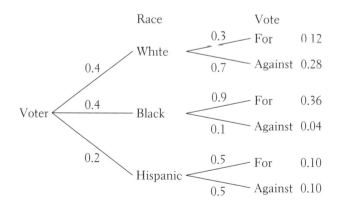

6.75 $P(\text{knows the answer} \mid \text{gives the correct answer}) = \frac{0.75}{0.80} = \frac{15}{16} = 0.9375$.

6.76 The event $\{Y < 1/2\}$ is the bottom half of the square, while $\{Y > X\}$ is the upper left triangle of the square. They overlap in a triangle with area 1/8, so

$$P(Y < \tfrac{1}{2} | Y > X) = \frac{P(Y < \tfrac{1}{2} \text{ and } Y > X)}{P(Y > X)} = \frac{1/8}{1/2} = \frac{1}{4}.$$

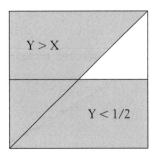

6.77 (a) P (switch bad) $= \frac{1000}{10,000} = 0.1$; P(switch OK) $= 1 - P$(switch bad) $= 0.9$. (b) Of the 9999 remaining switches, 999 are bad. P(second bad | first bad) $= \frac{999}{9999} \doteq 0.09991$. (c) Of the 9999 remaining switches, 1000 are bad. P(second bad | first good) $= \frac{1000}{9999} \doteq 0.10001$.

6.78 (a) There are 10 pairs. Just using initials: {(A,D), (A,J), (A,S), (A,R), (D,J), (D,S), (D,R), (J,S), (J,R), (S,R)}. (b) Each has probability $1/10 = 10\%$. (c) Julie is chosen in 4 of the 10 possible outcomes: $4/10 = 40\%$. (d) There are 3 pairs with neither Sam nor Roberto, so the probability is 3/10.

6.79 (a) P(Type AB) $= 1 - (0.45 + 0.40 + 0.11) = 0.04$.

(b) P(Type B or Type O) $= 0.11 + 0.45) = 0.56$.

(c) Assuming that the blood types for husband and wife are independent, P(Type B and Type A) $= (0.11)(0.40) = 0.044$.

(d) P(Type B and Type A) + P(Type A and Type B) $= (0.11)(0.40) + (0.40)(0.11) = 0.088$

(e) P(Husband Type O or Wife Type O) $= P$(Husband Type O) + P(Wife Type O) − P(Husband and Wife Type O) $= 0.45 + 0.45 - (0.45)^2 = 0.6975$.

6.80 (a) P(both have Type O) $= P$(Amer. has O) $\cdot P$(Chin. has O) $= (.45)(.35) = .1575$.

(b)P(both have same Type) $= (.45)(.35) + (.4)(.27) + (.11)(.26) + (.04)(.12) = .2989$.

6.81 (a) To find $P(A$ or $C)$, we would need to know $P(A$ and $C)$. (b) To find $P(A$ and $C)$, we would need to know $P(A$ or $C)$.

6.82

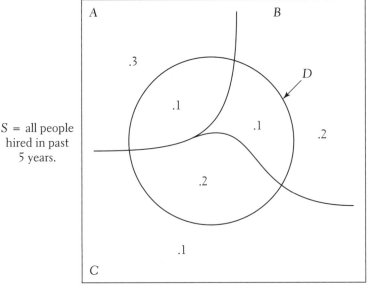

$P(O) = P(A \text{ and } D) + P(B \text{ and } D) + P(C \text{ and } D) = .1 + .1 + .2 = .4$

6.83 (a) $P(\text{firearm}) = \frac{18,940}{31,510} = 0.6011$.

(b) $P(\text{firearm} \mid \text{female}) = \frac{2,559}{6,095} = 0.4199$.

(c) $P(\text{female and firearm}) = \frac{2,559}{31,510} = 0.0812$.

(d) $P(\text{firearm} \mid \text{male}) = \frac{16,381}{25,415} = 0.6445$.

(e) $P(\text{male} \mid \text{firearm}) = \frac{16,381}{18,940} = 0.8649$.

6.84 Let H = adult belongs to health club, and let G = adult goes to club at least twice a week.
$P(G \text{ and } H) = P(H) \cdot P(G \mid H) = (.1)(.4) = .04$.

6.85 $P(B \mid A) = P(\text{both tosses have the same outcome} \mid H \text{ on first toss}) = 1/2 = 0.5$.

$P(B) = P(\text{both tosses have same outcome}) = 2/4 = 0.5$. Since $P(B \mid A) = P(B)$, events A and B are independent.

6.86

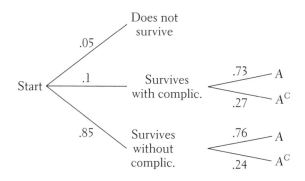

$P(A) = (.1)(.73) + (.85)(.76) = .719$. Surgery gives John a slightly larger chance of achieving his goal.

6.87 The response will be "no" with probability $0.35 = (0.5)(0.7)$. If the probability of plagiarism were 0.2, then $P(\text{student answers "no"}) = 0.4 = (0.05)(0.8)$. If 39% of students surveyed answered "no," then we estimate that $2 \times 39\% = 78\%$ have *not* plagiarized, so about 22% have plagiarized.

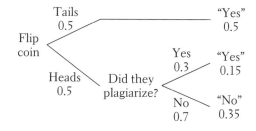

Random Variables

7

7.1 (a) *P*(less than 3) = *P*(1 or 2) = $\frac{2}{6} = \frac{1}{3}$. (b)–(c) Answers vary.

7.2 (a) BBB, BBG, BGB, GBB, GGB, GBG, BGG, GGG. Each has probability 1/8. (b) Three of the eight arrangements have two (and only two) girls, so *P*(*X* = 2) = 3/8 = 0.375. (c) See table.

Value of X	0	1	2	3
Probability	1/8	3/8	3/8	1/8

7.3 (a) 1%. (b) All probabilities are between 0 and 1; the probabilities add to 1. (c) *P*(*X* ≤ 3) = 0.48 + 0.38 + 0.08 = 1 − 0.01 − 0.05 = 0.94. (d) *P*(*X* < 3) = 0.48 + 0.38 = 0.86. (e) Write either *X* ≥ 4 or *X* > 3. The probability is 0.05 + 0.01 = 0.06. (f) Read two random digits from Table B. Here is the correspondence: 01 to 48 ⇔ Class 1, 49 to 86 ⇔ Class 2, 87 to 94 ⇔ Class 3, 95 to 99 ⇔ Class 4, and 00 ⇔ Class 5. Repeatedly generate 2 digit random numbers. The proportion of numbers in the range 01 to 94 will be an estimate of the required probability.

7.4

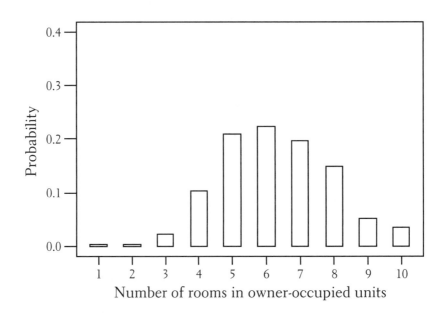

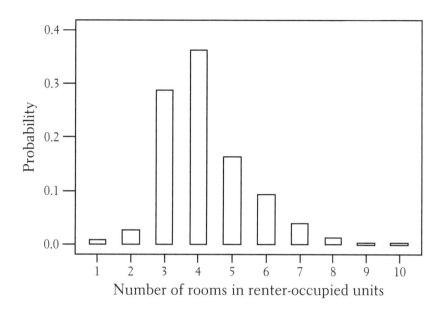

The rooms distribution is skewed to the right for renters and roughly symmetric for owners. This suggests that renter-occupied units tend, on the whole, to have fewer rooms than owner-occupied units.

7.5 (a) $\{X \geq 5\}$. $P(X \geq 5) = P(X = 5) + P(X = 6) + \ldots + P(X = 10) = 0.868$.

(b) $\{X > 5\}$ = the event that the unit has more than five rooms. $P(X > 5) = P(X = 6) + P(X = 7) + \ldots + P(X = 10) = 0.658$.

(c) A discrete random variable has a countable number of values, each of which has a distinct probability $(P(X = x))$. $P(X \geq 5)$ and $P(X > 5)$ are different because the first event contains the value $X = 5$ and the second does not.

7.6 (a) $P(0 \leq X \leq 0.4) = 0.4$.

(b) $P(0.4 \leq X \leq 1) = 0.6$.

(c) $P(0.3 \leq X \leq 0.5) = 0.2$.

(d) $P(0.3 < X < 0.5) = 0.2$.

(e) $P(0.226 \leq X \leq 0.713) = 0.713 - 0.226 = 0.487$.

(f) A continuous distribution assigns probability 0 to every individual outcome. In this case, the probabilities in (c) and (d) are the same because the events differ by 2 individual values, 0.3 and 0.5, each of which has probability 0.

7.7 (a) $P(X \leq 0.49) = 0.49$.

(b) $P(X \geq 0.27) = 0.73$.

(c) $P(0.27 < X < 1.27) = P(0.27 < X < 1) = 0.73$.

(d) $P(0.1 \leq X \leq 0.2 \text{ or } 0.8 \leq X \leq 0.9) = 0.1 + 0.1 = 0.2$.

(e) $P(\text{not } [0.3 \leq X < 0.8]) = 1 - 0.5 = 0.5$.

(f) $P(X = 0.5) = 0$.

7.8 (a) $P(\hat{p} \geq 0.45) = P\left(Z \geq \frac{0.45 - 0.4}{0.023}\right) = P(Z \geq 2.17) = 0.0150$.

(b) $P(\hat{p} < 0.35) = P(Z < -2.17) = 0.0150$.

(c) $P(0.35 \leq \hat{p} \leq 0.45) = P(-2.17 \leq Z \leq 2.17) = 0.9700$.

7.9 For a *sample* simulation of 400 observations from the N(0.4, 0.023) distribution, there were 0 observations less than 0.25, so the relative frequency is 0/400 = 0. The actual probability that $\hat{p} < 0.25$ is $P(Z < -6.52) \approx 3.5 \times 10^{-11}$, essentially 0.

7.10 (a) Both sets of probabilities sum to 1. (b) Both distributions are skewed to the right; however, the event $\{X = 1\}$ has a higher probability in the household distribution. This reflects the fact that a family must consist of two or more persons. Also, the events $\{X = 3\}$ and $\{X = 4\}$ have slightly higher probabilities in the family distribution, which may reflect the fact that families are more likely than households to have children living in the dwelling unit.

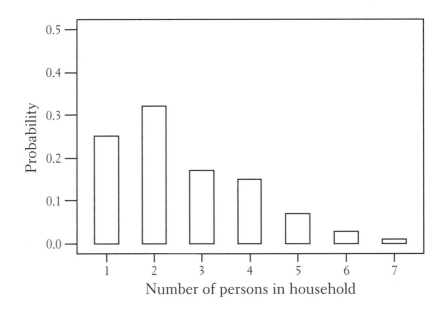

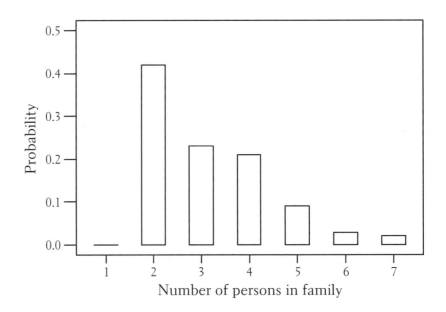

7.11 (a) $\{Y > 1\}$. $P(Y > 1) = P(Y = 2) + P(Y = 3) + \ldots + P(Y = 7) = 0.75$. Or, $P(Y > 1) = 1 - P(Y = 1) = 1 - .25 = .75$. (b) $P(2 < Y \le 4) = P(Y = 3) + P(Y = 4) = 0.32$. (c) $P(Y \ne 2) = 1 - P(Y = 2) = 0.68$.

7.12 (a) The probabilities sum to 1.

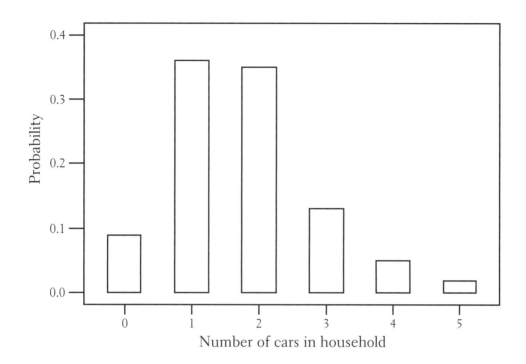

(b) $\{X \geq 1\}$ = the event that the household owns at least one car. $P(X \geq 1) = P(X = 1) + P(X = 2) + \ldots + P(X = 5) = 0.91$.

(c) $P(X > 2) = P(X = 3) + P(X = 4) + P(X = 5) = 0.20$. 20% of households own more cars than a two-car garage can hold.

7.13 (a) The 36 possible pairs of "up faces" are

$$\begin{array}{cccccc}
(1, 1) & (1, 2) & (1, 3) & (1, 4) & (1, 5) & (1, 6) \\
(2, 1) & (2, 2) & (2, 3) & (2, 4) & (2, 5) & (2, 6) \\
(3, 1) & (3, 2) & (3, 3) & (3, 4) & (3, 5) & (3, 6) \\
(4, 1) & (4, 2) & (4, 3) & (4, 4) & (4, 5) & (4, 6) \\
(5, 1) & (5, 2) & (5, 3) & (5, 4) & (5, 5) & (5, 6) \\
(6, 1) & (6, 2) & (6, 3) & (6, 4) & (6, 5) & (6, 6)
\end{array}$$

(b) Each pair must have probability 1/36.

(c) Let x = sum of up faces. Then

Sum	Outcomes	Probability
$x = 2$	(1, 1)	$p = 1/36$
$x = 3$	(1, 2) (2, 1)	$p = 2/36$
$x = 4$	(1, 3) (2, 2) (3, 1)	$p = 3/36$
$x = 5$	(1, 4) (2, 3) (3, 2) (4, 1)	$p = 4/36$
$x = 6$	(1, 5) (2, 4) (3, 3) (4, 2) (5, 1)	$p = 5/36$
$x = 7$	(1, 6) (2, 5) (3, 4) (4, 3) (5, 2) (6, 1)	$p = 6/36$
$x = 8$	(2, 6) (3, 5) (4, 4) (5, 3) (6, 2)	$p = 5/36$
$x = 9$	(3, 6) (4, 5) (5, 4) (6, 3)	$p = 4/36$
$x = 10$	(4, 6) (5, 5) (6, 4)	$p = 3/36$
$x = 11$	(5, 6) (6, 5)	$p = 2/36$
$x = 12$	(6, 6)	$p = 1/36$

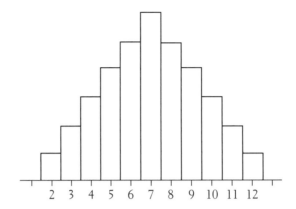

(d) $P(7 \text{ or } 11) = 6/36 + 2/36 = 8/36$ or $2/9$. (e) P (any sum other than 7) $= 1 - P(7) = 1 - 6/36 = 30/36 = 5/6$ by the complement rule.

7.14 Here is a table of the possible observations of Y that can occur when we roll one standard die and one "weird" die. As in Problem 7.13, there are 36 possible pairs of faces; however, a number of the pairs are identical to each other.

	1	2	3	4	5	6
0	1	2	3	4	5	6
0	1	2	3	4	5	6
0	1	2	3	4	5	6
6	7	8	9	10	11	12
6	7	8	9	10	11	12
6	7	8	9	10	11	12

The possible values of Y are 1, 2, 3, 4, ..., 12. Each value of Y has probability $\frac{3}{36} = \frac{1}{12}$.

7.15 (a) 75.2%. (b) All probabilities are between 0 and 1; the probabilities add to 1. (c) $P(X \geq 6) = 1 - 0.010 - 0.007 = 0.983$. (d) $P(X > 6) = 1 - 0.010 - 0.007 - 0.007 = 0.976$. (e) Either $X \geq 9$ or $X > 8$. The probability is $0.068 + 0.070 + 0.041 + 0.752 = 0.931$.

7.16 (a) $(0.6)(0.6)(0.4) = 0.144$. (b) The possible combinations are SSS, SSO, SOS, OSS, SOO, OSO, OOS, OOO (S = support, O = oppose). $P(SSS) = 0.6^3 = 0.216$, $P(SSO) = P(SOS) = P(OSS) = (0.6^2)(0.4) = 0.144$, $P(SOO) = P(OSO) = P(OOS) = (0.6)(0.4^2) = 0.096$, and $P(OOO) = 0.4^3 = 0.064$. (c) The distribution is given in the table: The probabilities are found by adding the probabilities from (b), noting that (e.g.) $P(X = 1) = P(SSO \text{ or } SOS \text{ or } OSS)$. (d) Write either $X \geq 2$ or $X > 1$. The probability is $0.288 + 0.064 = 0.352$.

Value of X	0	1	2	3
Probability	0.216	0.432	0.288	0.064

7.17 (a) The height should be $\frac{1}{2}$, since the area under the curve must be 1. The density curve is below. (b) $P(y \leq 1) = \frac{1}{2}$, (c) $P(0.5 < y < 1.3) = 0.4$. (d) $P(y \geq 0.8) = 0.6$.

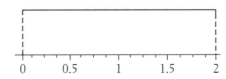

7.18 (a) The area of a triangle is $\frac{1}{2}bh = \frac{1}{2}(2)(1) = 1$. (b) $P(Y < 1) = 0.5$. (c) $P(Y < 0.5) = 0.125$.

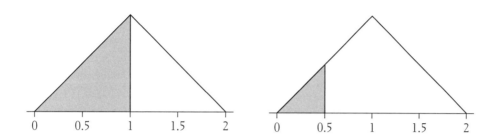

7.19 The resulting histogram should *approximately* resemble the triangular density curve of Figure 7.8, with any deviations or irregularities depending upon the specific random numbers generated.

7.20 (a) $P(\hat{p} \geq 0.16) = P(Z \geq \frac{0.16 - 0.15}{0.0092}) = P(Z \geq 1.09) = 0.1379$. (b) $P(0.14 \leq \hat{p} \leq 0.16) = P(-1.09 \leq Z \leq 1.09) = 0.7242$.

7.21 In this case, we will simulate 500 observations from the N(.15, .0092) distribution. The required TI-83 commands are as follows:

$$ClrList\ L_1$$
$$randNorm\ (.15, .0092, 500) \rightarrow L_1$$
$$sortA(L_1)$$

Scrolling through the 500 simulated observations, we can determine the relative frequency of observations that are at least .16 by using the complement rule. For a sample simulation, there are 435 observations less than .16, thus the desired relative frequency is $1 - 435/500 = 65/500 = .13$. The actual probability that $p \geq .16$ is .1385. 500 observations yield a reasonably close approximation.

7.22 $\mu = (0)(0.10) + (1)(0.15) + (2)(0.30) + (3)(0.30) + (4)(0.15) = 2.25$.

7.23 Owner-occupied units: $\mu = (1)(.003) + (2)(.002) + (3)(.023) + (4)(.104) + (5)(.210) + (6)(.224) + (7)(.197) + (8)(.149) + (9)(.053) + (10)(.035) = 6.284$.
Renter-occupied units: $\mu = (1)(.008) + (2)(.027) + (3)(.287) + (4)(.363) + (5)(.164) + (6)(.093) + (7)(.039) + (8)(.013) + (9)(.003) + (10)(.003) = 4.187$.
The larger value of μ for owner-occupied units reflects the fact that the owner distribution was symmetric, rather than skewed to the right, as was the case with the renter distribution. The "center" of the owner distribution is roughly at the central peak class, 6, whereas the "center" of the renter distribution is roughly at the class 4.

7.24 (a) If your number is *abc*, then of the 1000 three-digit numbers, there are six—*abc, acb, bac, bca, cab, cba*—for which you will win the box. Therefore, we win nothing with probability $\frac{994}{1000} = 0.994$ and win \$83.33 with probability $\frac{6}{1000} = 0.006$.

   (b) The expected payoff on a \$1 bet is $\mu = (\$0)(0.994) + (\$83.33)(0.006) = \$0.50$.

   (c) The casino keeps 50 cents from each dollar bet in the long run, since the expected payoff = 50 cents.

7.25 (a) The payoff is either \$0 or \$3; see table on next page. (b) For each \$1 bet, $\mu_x = (\$0)(0.75) + (\$3)(0.25) = \$0.75$. (c) The casino makes 25 cents for every dollar bet (in the long run).

Value of X	0	3
Probability	0.75	0.25

7.26 In 7.22, we had $\mu = 2.25$, so $\sigma_x^2 = (0 - 2.25)^2(0.10) + (1 - 2.25)^2(0.15) + (2 - 2.25)^2(0.30) + (3 - 2.25)^2(0.30) + (4 - 2.25)^2(0.15) = 1.3875$, and $\sigma_x = \sqrt{1.3875} = 1.178$.

7.27 Mean:

Household size: $\mu = (1)(.25) + (2)(.32) + (3)(.17) + (4)(.15) + (5)(.07) + (6)(.03) + (7)(.01) = 2.6$.

Family size: $\mu = (1)(0) + (2)(.42) + (3)(.23) + (4)(.21) + (5)(.09) + (6)(.03) + (7)(.02) = 3.14$.

Standard deviation:

Household size: $\sigma^2 = (1 - 2.6)^2(.25) + (2 - 2.6)^2(.32) + (3 - 2.6)^2(.17) + (4 - 2.6)^2(.15) + (5 - 2.6)^2(.07) + (6 - 2.6)^2(.03) + (7 - 2.6)^2(.01) = 2.02$, and $\sigma = \sqrt{2.02} = 1.421$.

Family size: $\sigma^2 = (1 - 3.14)^2(0) + (2 - 3.14)^2(.42) + (3 - 3.14)^2(.23) + (4 - 3.14)^2(.21) + (5 - 3.14)^2(.09) + (6 - 3.14)^2(.03) + (7 - 3.14)^2(.02) = 1.5604$, and $\sigma = \sqrt{1.5604} = 1.249$.

The family distribution has a slightly larger mean than the household distribution, reflecting the fact that the family distribution assigns more "weight" (probability) to the values 3 and 4. The two standard deviations are roughly equivalent; the household standard deviation may be larger because of the fact that it assigns a nonzero probability to the value 1.

7.28 We would expect the owner distribution to have a wider spread than the renter distribution. The central "peak" of the owner distribution is more spread out than the left-hand "peak" of the renter distribution, and as a result the average distance between a value and the mean is slightly larger in the owner case.

Owner-occupied units: $\sigma^2 = (1 - 6.284)^2(.003) + (2 - 6.284)^2(.002) + (3 - 6.284)^2(.023) + (4 - 6.284)^2(.104) + (5 - 6.284)^2(.210) + (6 - 6.284)^2(.224) + (7 - 6.284)^2(.197) + (8 - 6.284)^2(.149) + (9 - 6.284)^2(.053) + (10 - 6.284)^2(.035) = 2.689344$, and $\sigma = \sqrt{2.689344} = 1.64$.

Renter-occupied units: $\mu = (1 - 4.187)^2(.008) + (2 - 4.187)^2(.027) + (3 - 4.187)^2(.287) + (4 - 4.187)^2(.363) + (5 - 4.187)^2(.164) + (6 - 4.187)^2(.093) + (7 - 4.187)^2(.039) + (8 - 4.187)^2(.013) + (9 - 4.187)^2(.003) + (10 - 4.187)^2(.003) = 1.710031$, and $\sigma = \sqrt{1.710031} = 1.308$.

7.29 (a) $\mu_x = (0)(0.03) + (1)(0.16) + (2)(0.30) + (3)(0.23) + (4)(0.17) + (5)(0.11) = 2.68$. $\sigma_x^2 = (0 - 2.68)^2(0.03) + (1 - 2.68)^2(0.16) + (2 - 2.68)^2(0.30) + (3 - 2.68)^2(0.23) + (4 - 2.68)^2(0.17) + (5 - 2.68)^2(0.11) = 1.7176$, and $\sigma_x = \sqrt{1.7176} = 1.3106$.

(b) To simulate (say) 500 observations of x, using the T1-83, we will first simulate 500 random integers between 1 and 100 by using the command:

$$\text{randInt}(1,100,500) \rightarrow L_1$$

The command sortA(L_1) sorts these random observations in increasing order. We now identify 500 observations of x as follows:

Integers		correspond to	
	1 to 3		$x = 0$
	4 to 19		$x = 1$
	20 to 49		$x = 2$
	50 to 72		$x = 3$
	73 to 89		$x = 4$
	90 to 100		$x = 5$

For a sample run of the simulation, we obtained

12	observations of	$x = 0$
86		$x = 1$
155		$x = 2$
118		$x = 3$
75		$x = 4$
54		$x = 5$

These data yield a sample mean and standard deviation of $\bar{x} = 2.64$, $s = 1.292$, very close to μ and σ.

7.30 The graph for $x_{max} = 10$ displays visible variation for the first ten values of x, whereas the graph for $x_{max} = 100$ gets closer and closer to $\mu = 64.5$ as x increases. This illustrates that the larger the sample size (represented by the integers 1, 2, 3, ... in L_1), the closer the sample means x get to the population mean $\mu = 64.5$. (In other words, this exercise illustrates the law of large numbers in a graphical manner.)

7.31 Below is the probability distribution for L, the length of the longest run of heads or tails. $P(\text{You win}) = P(\text{run of 1 or 2}) = \frac{89}{512} = 0.1738$, so the expected outcome is $\mu = (\$2)(0.1738) + (-\$1)(0.8262) = -\$0.4785$. On the average, you will lose about 48 cents each time you play. (Simulated results should be close to this exact result; how close depends on how many trials are used.)

Value of L	1	2	3	4	5	6	7	8	9	10
Probability	$\frac{1}{512}$	$\frac{88}{512}$	$\frac{185}{512}$	$\frac{127}{512}$	$\frac{63}{512}$	$\frac{28}{512}$	$\frac{12}{512}$	$\frac{5}{512}$	$\frac{2}{512}$	$\frac{1}{512}$

7.32 (a) The wheel is not affected by its past outcomes—it has no memory; outcomes are independent. So on any one spin, black and red remain equally likely.

(b) Removing a card changes the composition of the remaining deck, so successive draws are not independent. If you hold 5 red cards, the deck now contains 5 fewer red cards, so your chance of another red decreases.

7.33 No: Assuming all "at-bat"s are independent of each other, the 35% figure only applies to the "long run" of the season, not to "short runs."

7.34 (a) Independent: Weather conditions a year apart should be independent. (b) Not independent: Weather patterns tend to persist for several days; today's weather tells us something about tomorrow's. (c) Not independent: The two locations are very close together, and would likely have similar weather conditions.

7.35 (a) Dependent: since the cards are being drawn from the deck without replacement, the nature of the third card (and thus the value of Y) will depend upon the nature of the first two cards that were drawn (which determine the value of X).

(b) Independent: X relates to the outcome of the first roll, Y to the outcome of the second roll, and individual dice rolls are independent (the dice have no memory).

7.36 The total mean is $40 + 5 + 25 = 70$ minutes.

7.37 (a) The total mean is $11 + 20 = 31$ seconds. (b) No: Changing the standard deviations does not affect the means. (c) No: The total mean does not depend on dependence or independence of the two variables.

7.38 Assuming that the two times are independent, the total variance is $\sigma_{total}^2 = \sigma_{pos}^2 + \sigma_{att}^2 = 2^2 + 4^2 = 20$, so $\sigma_{total} = \sqrt{20} = 4.472$ seconds. Assuming that the two times are dependent with correlation 0.3, the total variance is $\sigma_{total}^2 = \sigma_{pos}^2 + \sigma_{att}^2 + 2\rho\sigma_{pos}\sigma_{att} = 2^2 + 4^2 + 2(0.3)(2)(4) = 24.8$, so $\sigma_{total} = \sqrt{24.8} = 4.98$ seconds. The positive correlation of 0.3 indicates that the two times have some tendency to either increase together or decrease together, which increases the variability of their sum.

7.39 Since the two times are independent, the total variance is $\sigma_{total}^2 = \sigma_{first}^2 + \sigma_{second}^2 = 2^2 + 1^2 = 5$, so $\sigma_{total} = \sqrt{5} = 2.236$ minutes.

7.40 (a) $\sigma_Y^2 = (300 - 445)^2(0.4) + (500 - 445)^2(0.5) + (750 - 455)^2(0.1) = 19.225$ and $\sigma_Y = 138.65$ units. (b) $\sigma_{X+Y}^2 = \sigma_X^2 + \sigma_Y^2 = 7{,}800{,}000 + 19{,}225 = 7{,}819{,}225$, so $\sigma_{X+Y} = 2796.29$ units. (c) $\sigma_Z^2 = \sigma_{2000X}^2 + \sigma_{3500Y}^2 = (2000)^2\sigma_X^2 + (3500)^2\sigma_Y^2$, so $\sigma_Z = \$5{,}606{,}738$.

7.41 If F and L are their respective scores, then $F - L$ has a $N(0, \sqrt{2^2 + 2^2}) = N(0, 2\sqrt{2})$ distribution, so $P(|F - L| > 5) = P(|Z| > 1.7678) = 0.0771$ (table value: 0.0768).

7.42 (a) Let X = the value of the stock after two days. The possible combinations of gains and losses on two days are presented in the table below, together with the calculation of the corresponding values of X.

1st day	2nd day	Value of X
Gain 30%	Gain 30%	$1000 + (.3)(1000) = 1300$
		$1300 + (.3)(1300) = 1690$
Gain 30%	Lose 25%	$1000 + (.3)(1000) = 1300$
		$1300 - (.25)(1300) = 975$
Lose 25%	Gain 30%	$1000 - (.25)(1000) = 750$
		$750 + (.3)(750) = 975$
Lose 25%	Lose 25%	$1000 - (.25)(1000) = 750$
		$750 - (.25)(750) = 562.50$

Since the returns on the two days are independent and $P(\text{gain } 30\%) = P(\text{lose } 25\%) = 0.5$, the probability of each of these combinations is $(.5)(.5) = .25$. The probability distribution of X is therefore

x	1690	975	562.5
$P(X = x)$	0.25	0.5	0.25

The probability that the stock is worth more than $\$1000 = P(X = 1690) = 0.25$.
(b) $\mu = (1690)(.25) + (975)(.5) + (562.5)(.25) = 1050.625$, or approximately $\$1051$.

7.43 The probability distribution of digits under Benford's Law reflects *long-term* behavior. For each digit v, $P(V = v) \approx$ the long-term relative frequency of v, with the accuracy of the approximation improving as the number of observations increases. We would expect a large number of items to reflect Benford's Law more accurately than a small number of items.

7.44 (a) First die: $\mu = (1)(\frac{1}{6}) + (3)(\frac{1}{6}) + (4)(\frac{1}{6}) + (5)(\frac{1}{6}) + (6)(\frac{1}{6}) + (8)(\frac{1}{6}) = 4.5$.
Second die: $\mu = (1)(\frac{1}{6}) + (2)(\frac{1}{3}) + (3)(\frac{1}{3}) + (4)(\frac{1}{6}) = 2.5$.

(b) The table on the facing page gives the distribution of X = sum of spots for the two dice. Each of the 36 observations in the table has probability $\frac{1}{36}$.

	1	3	4	5	6	8
1	2	4	5	6	7	9
2	3	5	6	7	8	10
2	3	5	6	7	8	10
3	4	6	7	8	9	11
3	4	6	7	8	9	11
4	5	7	8	9	10	12

The probability distribution of X is:

x	2	3	4	5	6	7	8	9	10	11	12
$P(X = x)$	$\frac{1}{36}$	$\frac{1}{18}$	$\frac{1}{12}$	$\frac{1}{9}$	$\frac{5}{36}$	$\frac{1}{6}$	$\frac{5}{36}$	$\frac{1}{9}$	$\frac{1}{12}$	$\frac{1}{18}$	$\frac{1}{36}$

(c) $\mu = (2)(\frac{1}{36}) + (3)(\frac{1}{18}) + (4)(\frac{1}{12}) + (5)(\frac{1}{9}) + (6)(\frac{5}{36}) + (7)(\frac{1}{6}) + (8)(\frac{5}{36}) + (9)(\frac{1}{9}) + (10)(\frac{1}{12}) + (11)(\frac{1}{18}) + (12)(\frac{1}{36}) = 7$.
Using addition rule for means: μ = mean from first die + mean from second die = 4.5 + 2.5 = 7.

7.45 (a) Randomly selected students would presumably be unrelated. (b) $\mu_{f-m} = \mu_f - \mu_m = 120 - 105 = 15$. $\sigma_{f-m}^2 = \sigma_f^2 + \sigma_m^2 = 28^2 + 35^2 = 2009$, so $\sigma_{f-m} = 44.82$. (c) Knowing only the mean and standard deviation, we cannot find that probability (unless we assume that the distribution is normal). Many different distributions can have the same mean and standard deviation.

7.46 (a) $\mu_x = 550°$ Celsius; $\sigma_x^2 = 32.5$, so $\sigma_x = 5.701°C$. (b) Mean: $0°C$; standard deviation: $5.701°C$. (c) $\mu_y = \frac{9}{5}\mu_x + 32 = 1022°F$, and $\sigma_y = \frac{9}{5}\sigma_x = 10.26°F$.

7.47 Read two-digit random numbers. Establish the correspondence 01 to 10 $\Leftrightarrow$ 540°, 11 to 35 $\Leftrightarrow$ 545°, 36 to 65 $\Leftrightarrow$ 550°, 66 to 90 $\Leftrightarrow$ 555°, and 91 to 99, 00 $\Leftrightarrow$ 560°. Repeat many times, and record the corresponding temperatures. Average the temperatures to approximate μ; find the standard deviations of the temperatures to approximate σ.

7.48 (a) The machine that makes the caps and the machine that applies the torque are not the same. (b) T (torque) is $N(7, 0.9)$ and S (cap strength) is $N(10, 1.2)$, so $T - S$ is $N(-3, \sqrt{0.9^2 + 1.2^2}) = N(-3, 1.5)$. Then $P(T > S) = P(T - S > 0) = P(Z > 2) = 0.0228$.

7.49 (a) Yes: This is always true; it does not depend on independence. (b) No: It is not reasonable to believe that X and Y are independent.

7.50 (a) $R_1 + R_2$ is normal with mean $100+250 = 350\Omega$ and s.d. $\sqrt{2.5^2 + 2.8^2} \doteq 3.7537\Omega$. (b) $P(345 \le R_1 + R_2 \le 355) = P(-1.3320 \le Z \le 1.3320) = 0.8172$ (table value: 0.8164).

7.51 The monthly return on a portfolio of 80% Magellan and 20% Japan can be written as $0.8W + 0.2Y$. The mean return is $\mu_{0.8W+0.2Y} = 0.8\mu_W + 0.2\mu_Y = (0.8)(1.14) + (0.2)(1.59) = 1.23\%$, which is higher than μ_W. The variance of return is $\sigma_{0.8W+0.2Y}^2 = (0.8)^2\sigma_W^2 + (0.2)^2\sigma_Y^2 + 2\rho_{WY}(0.8\sigma_W)(0.2\sigma_Y) = (0.64)(4.64)^2 + (0.04)(6.75)^2 + 2(0.54)(0.8)(4.64)(0.2)(6.75) = 21.01354$. The standard deviation of return is $\sigma_{0.8W+0.2Y} = \sqrt{21.01354} = 4.584\%$, which is lower than σ_W.

7.52 Assuming that $\rho_{WY} = 0$, the variance of return is $\sigma_{0.8W+0.2Y}^2 = (0.8)^2\sigma_W^2 + (0.2)^2\sigma_Y^2 = (0.64)(4.64)^2 + (0.04)(6.75)^2 = 15.601444$, and the standard deviation is $\sigma_{0.8W+0.2Y} = \sqrt{15.601444} = 3.95\%$. This is smaller than the result 4.584% from Problem 7.51. The mean return is not affected by the zero correlation, since it only depends upon the means of the individual returns.

7.53 The monthly return on a portfolio of 60% Magellan, 20% Real Estate, and 20% Japan can be expressed as $0.6W + 0.2X + 0.2Y$. The mean return is $\mu_{0.6W+0.2X+0.2Y} = 0.6\mu_W + 0.2\mu_X + 0.2\mu_Y = (0.6)(1.14) + (0.2)(0.16) + (0.2)(1.59) = 1.034\%$. The variance of return is $\sigma^2_{0.6W+0.2X+0.2Y} = (0.6)^2\sigma^2_W + (0.2)^2\sigma^2_X + (0.2)^2\sigma^2_Y + 2\rho_{WX}(0.6\sigma_W)(0.2\sigma_X) + 2\rho_{WY}(0.6\sigma_W)(0.2\sigma_Y) + 2\rho_{XY}(0.2\sigma_X)(0.2\sigma_Y) = (0.36)(4.64)^2 + (0.04)(3.61)^2 + (0.04)(6.75)^2 + 2(0.19)(0.6)(4.64)(0.2)(3.61) + 2(0.54)(0.6)(4.64)(0.2)(6.75) + 2(-0.17)(0.2)(3.61)(0.2)(6.75) = 14.58593224$. The standard deviation of return is $\sqrt{14.58593224} = 3.82\%$.

7.54 (a) Writing (x, y), where x is Ann's choice and y is Bob's choice, the sample space has 16 elements:

(A,A)	(A,B)	(A,C)	(A,D)	(B,A)	(B,B)	(B,C)	(B,D)
0	2	−3	0	−2	0	0	3

(C,A)	(C,B)	(C,C)	(C,D)	(D,A)	(D,B)	(D,C)	(D,D)
3	0	0	−4	0	−3	4	0

(b) The value of X is written below each entry in the table.
(c) Below.
(d) The mean is 0, so the game is fair. The variance is 4.75, so $\sigma_X \doteq 2.1794$.

Value of X	−4	−3	−2	0	2	3	4
Probability	$\frac{1}{16}$	$\frac{2}{16}$	$\frac{1}{16}$	$\frac{8}{16}$	$\frac{1}{16}$	$\frac{2}{16}$	$\frac{1}{16}$

7.55 The missing probability is 0.99058 (so that the sum is 1). This gives mean earnings $\mu_x = \$303.3525$.

7.56 The mean μ of the company's "winnings" (premiums) and their "losses" (insurance claims) is positive. Even though the company will lose a large amount of money on a small number of policyholders who die, it will gain a small amount on the majority. The law of large numbers says that the average "winnings" minus "losses" should be close to μ, and overall the company will almost certainly show a profit.

7.57 $\sigma^2_X = 94{,}236{,}826.64$, so that $\sigma_X = \$9707.57$.

7.58 (a) $\mu_Z = \frac{1}{2}\mu_T = \mu_X = \$303.3525$. $\sigma_Z = \sqrt{\frac{1}{4}\sigma^2_X + \frac{1}{4}\sigma^2_Y} = \sqrt{\frac{1}{2}\sigma^2_X} = \$6864.29$. (b) With this new definition of Z: $\mu_Z = \mu_X = \$303.3525$ (unchanged). $\sigma_Z = \sqrt{\frac{1}{4}\sigma^2_X} = \frac{1}{2}\sigma_X = \$4853.78$ (smaller by a factor of $1/\sqrt{2}$).

7.59 $X - Y$ is $N(0, \sqrt{0.3^2 + 0.3^2}) \doteq N(0, 0.4243)$, so $P(|X - Y| \geq 0.8) = P(|Z| \geq 1.8856) = 1 - P(|Z| \leq 1.8856) = 0.0593$ (table value: 0.0588).

7.60 (a) $\mu_X = (1)(0.1) + (1.5)(0.2) + (2)(0.4) + (4)(0.2) + (10)(0.1) = 3$ million dollars. $\sigma^2_X = (4)(0.1) + (2.25)(0.2) + (1)(0.4) + (1)(0.2) + (49)(0.1) = 503.375$. so $\sigma_X \doteq 22.436$ million dollars.
   (b) $\mu_Y = 0.9\mu_X - 0.2 = 2.5$ million dollars, and $\sigma_Y = 0.9\sigma_X \doteq 20.192$ million dollars.

7.61 (a) $\mu_{Y-X} = \mu_Y - \mu_X = 2.001 - 2.000 = 0.001$ g. $\sigma_{Y-X}^2 = \sigma_Y^2 + \sigma_X^2 = 0.002^2 + 0.001^2 = 0.000005$, so $\sigma_{Y-X} = 0.002236$ g.
   (b) $\mu_Z = \frac{1}{2}\mu_X + \frac{1}{2}\mu_Y = 2.0005$ g. $\sigma^2_Z = \frac{1}{4}\sigma^2_X + \frac{1}{4}\sigma^2_Y = 0.00000125$, so $\sigma_Z = 0.001118$ g. Z is slightly more variable than Y, since $\sigma_Y < \sigma_Z$.

7.62 (a) To do one repetition, start at any point in Table B and begin reading digits. As in Example 5.24, let the digits 0, 1, 2, 3, 4 = girl and 5, 6, 7, 8, 9 = boy, and read a string of digits until a "0 to 4" (girl) appears or until four consecutive "5 to 9"s (boys) have appeared, whichever comes first. Then let the observation of x = number of children for this repetition = the number of digits in the string you have read. Repeat this procedure 25 times to obtain your 25 observations.

(b) The possible outcomes and their corresponding values of x = number of children are as follows:

	Outcome	
$x = 1$	G	(first child is a girl)
$x = 2$	BG	(second child is a girl)
$x = 3$	BBG	(third child is a girl)
$x = 4$	BBBG, BBBB	(four children)

Using the facts that births are independent, the fact that B and G are equally likely to occur on any one birth, and the multiplication rule for independent events, we find that

$$P(x = 1) = 1/2$$
$$P(x = 2) = (1/2)(1/2) = 1/4$$
$$P(x = 3) = (1/2)(1/2)(1/2) = 1/8$$
$$P(x = 4) = (1/2)(1/2)(1/2)(1/2) + (1/2)(1/2)(1/2)(1/2)$$
$$= 1/16 + 1/16 = 1/8$$

The probability distribution of x is therefore:

x_i	1	2	3	4
p_i	1/2	1/4	1/8	1/8

(c) $\mu_x = \Sigma x_i p_i$
$$= (1)(1/2) + (2)(1/4) + (3)(1/8) + (4)(1/8)$$
$$= 1/2 + 1/2 + 3/8 + 1/2$$
$$= 1.875$$

7.63 (a) A single random digit simulates each toss, with (say) odd = heads and even = tails. The first round is two digits, with two odds a win; if you don't win, look at two more digits, again with two odds a win.

(b) The probability of winning is $\frac{1}{4} + (\frac{3}{4})(\frac{1}{4}) = \frac{7}{16}$, so the expected value is $(\$1)(\frac{7}{16}) + (-\$1)(\frac{9}{16}) = -\frac{2}{16} = -\$0.125$.

7.64 $\mu_X = (\mu - \sigma)(0.5) + (\mu + \sigma)(0.5) = \mu$, and $\sigma_X = \sigma$ since $\sigma_X^2 = [\mu - (\mu - \sigma)]^2(0.5) = [\mu - (\mu + \sigma)]^2(0.5) = \sigma^2(0.5) + \sigma^2(0.5) = \sigma^2$.

7.65 By the general addition rule for variances, $\sigma_{X+Y}^2 = \sigma_X^2 + \sigma_Y^2 + 2\rho\sigma_X\sigma_Y = \sigma_X^2 + \sigma_Y^2 + 2(1)\sigma_X\sigma_Y = \sigma_X^2 + \sigma_Y^2 + 2\sigma_X\sigma_Y = (\sigma_X + \sigma_Y)^2$. Taking square roots yields $\sigma_{X+Y} = \sigma_X + \sigma_Y$.

7.66 (a) Two standard deviations: $d_1 = 2(0.002) = 0.004$ and $d_2 = 2(0.001) = 0.002$.
(b) $\sigma_{X+Y+Z} = \sqrt{0.002^2 + 0.001^2 + 0.001^2} \doteq 0.002449$, so $d \doteq 0.005$—considerably *less* than $d_1 + 2d_2 = 0.008$. The engineer was incorrect.

7.67 We want to find a, b such that $\mu_Y = a + b\mu_X = 0$ and $\sigma_Y^2 = b^2\sigma_X^2 = 1$. Substituting $\mu_X = 1400$ and $\sigma_X = 20$, we have $a + b(1400) = a + 1400b = 0$ and $b^2(20)^2 = 400b^2 = 1$. Solving the second equation for b yields $b = \frac{1}{20}$. Substituting this value into the first equation and solving for a yields $a = -1400(\frac{1}{20}) = -70$.

7.68 (a) The possible values of X are 3, 4, 5, 6, ..., 18 (all positive integers between 3 and 18).

(b) We get a sum of 5 if and only if either one die shows 3 and the other two dice show 1's or two dice show 2's and the third shows 1. Each of these arrangements can occur in three ways, thus the event $\{X = 5\}$ contains 6 outcomes. The total number of possible outcomes when three dice are rolled is $6 \times 6 \times 6 = 216$. $P(X = 5) = \frac{6}{216} = \frac{1}{36}$.

(c) For each X_i, the mean $\mu_i = 3.5$ and the variance $\sigma_i^2 = 2.917$. The mean of the sum $X = \mu_1 + \mu_2 + \mu_3 = 3(3.5) = 10.5$. The X_i are independent, so the variance of the sum $X = \sigma_1^2 + \sigma_2^2 + \sigma_3^2 = 3(2.917) = 8.751$, and the standard deviation of $X = \sqrt{8.751} = 2.958$.

8

The Binomial and Geometric Distributions

8.1 (a) No: There is no fixed n (i.e., there is no definite upper limit on the number of defects). (b) Yes: It is reasonable to believe that all responses are independent (ignoring any "peer pressure"), and all have the same probability of saying "yes" since they are randomly chosen from the population. Also, a "large city" will have a population over 1000 (10 times as big as the sample). (c) Yes: In a "Pick 3" game, Joe's chance of winning the lottery is the same every week, so assuming that a year consists of 52 weeks (observations), this would be binomial.

8.2 (a) Yes: It is reasonable to assume that the results for the 50 students are independent, and each has the same chance of passing. (b) No: Since the student receives instruction after incorrect answers, her probability of success is likely to increase. (e) No: Temperature may affect the outcome of the test.

8.3

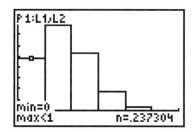 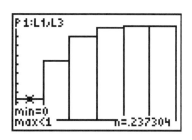

(a) .2637. (b) The binomial probabilities for $x = 0, \ldots, 5$ are: .2373, .3955, .2637, .0879, .0146, .0010. (e) The cumulative probabilities for $x = 0, \ldots, 5$ are: .2373, .6328, .8965, .9844, .9990, 1. Compared with Corinne's cdf histogram, the bars in this histogram get taller, sooner. Both peak at 1 on the extreme right.

8.4 Let X = the number of correct answers. X is binomial with $n = 50$, $p = 0.5$.

    (a) $P(X \geq 25) = 1 - P(X \leq 24) = 1 - $ binomcdf $(50, .5, 24) = 1 - .444 = .556.$

    (b) $P(X \geq 30) = 1 - P(X \leq 29) = 1 - $ binomcdf $(50, .5, 29) = 1 - .899 = .101.$

    (c) $P(X \geq 32) = 1 - P(X \leq 31) = 1 - $ binomcdf $(50, .5, 31) = 1 - .968 = .032.$

8.5 (a) Let X = the number of correct answers. X is binomial with $n = 10$, $p = 0.25$. The probability of at least one correct answer is $P(X \geq 1) = 1 - P(X = 0) = 1 - $ binompdf $(10, .25, 0) = 1 - .056 = .944.$

(b) Let X = the number of correct answers. We can write $X = X_1 + X_2 + X_3$, where X_i = the number of correct answers on question i. (Note that the only possible values of X_i are 0 and 1, with 0 representing an incorrect answer and 1 a correct answer.) The probability of at least one correct answer is $P(X \geq 1) = 1 - P(X = 0) = 1 - [P(X_1 = 0)P(X_2 = 0) P(X_3 = 0)]$ (since the X_i are independent) $= 1 - \left(\frac{2}{3}\right)\left(\frac{3}{4}\right)\left(\frac{4}{5}\right) = 1 - \frac{24}{60} = 0.6$.

8.6 (a) Yes, if the 100 children are randomly selected, it is extremely likely that the result for one child will not be influenced by the result for any other child (e.g., the children are siblings). "Success" in this context means having an incarcerated parent. $n = 100$, since 100 children are selected, and $p = 0.02$.

(b) $P(X = 0) =$ the probability of none of the 100 selected children having an incarcerated parent. $P(X = 0) = \mathtt{binompdf}$ $(100, .02, 0) = .133$. $P(X = 1) = \mathtt{binompdf}$ $(100, .02, 1) = .271$.

(c) $P(X \geq 2) = 1 - P(X \leq 1) = 1 - \mathtt{binomcdf}$ $(100, .02, 1) = 1 - .403 = .597$. Alternatively, by the addition rule for mutually exclusive events, $P(X \geq 2) = 1 - (P(X = 0) + P(X = 1)) = 1 - (.133 + .271) = 1 - .404 = .596$. (The difference between answers is due to roundoff error.)

8.7 Let $X =$ the number of players out of 20 who graduate. $P(X = 11) = \mathtt{binompdf}$ $(20, .8, 11) = .0074$.

8.8 (a) $n = 10$ and $p = 0.25$. (b) $\binom{10}{2} (0.25)^2(0.75)^8 = 0.28157$. (c) $P(X \leq 2) = \binom{10}{0}(0.25)^0(0.75)^{10} + \ldots + \binom{10}{2}(0.25)^2(0.75)^8 = 0.52559$.

8.9 $P(X = 3) = \binom{5}{3}(.25)^3(.75)^2 = (10)(.25)^3(.75)^2 = .088$.

8.10 Let $X =$ the number of broccoli plants that you lose. X is $B(10, .05)$.

$$P(X \leq 1) = P(X = 0) + P(X = 1) = \binom{10}{0} (.05)^0(.95)^{10} + \binom{10}{1} (.05)^1(.95)^9 = (.95)^{10} + (10)(.05)(.95)^9 = .914.$$

8.11 Let $X =$ the number of children with blood type O. X is $B(5, .25)$.

$$P(X \geq 1) = 1 - P(X = 0) = 1 - \binom{5}{0} (.25)^0(.75)^5 = 1 - (.75)^5 = .763.$$

8.12 Probability that all 20 graduate: $P(X = 20) = \binom{20}{20} (.8)^{20}(.2)^0 = (.8)^{20} = .0115$.

Probability that not all 20 graduate: $P(X < 20) = 1 - P(X = 20) = .9885$.

8.13 (a) $\binom{15}{3} (0.3)^3(0.7)^{12} = 0.17004$. (b) $\binom{15}{0} (0.3)^0(0.7)^{15} = .00475$.

8.14 Let $X =$ the number of free throws that Corinne makes. X is $B(12, .7)$.

$$P(X = 7) = \binom{12}{7} (.75)^7(.25)^5 = (792)(.75)^7(.25)^5 = .1032.$$

8.15 (a) $np = 1500$, $n(1 - p) = 1000$; both values are greater than or equal to 10. (b) Let $X =$ the number of people in the sample who find shopping frustrating. X is $B(2500, .6)$. Then $P(X \geq 1520) = 1 - P(X \leq 1519) = 1 - \mathtt{binomcdf}$ $(2500, .6, 1519) = 1 - .7868609113 = .2131390887$, which rounds to .2131. The probability correct to six decimal places is .213139. (c) $P(X \leq 1468) = \mathtt{binomcdf}$ $(2500, .6, 1468) = .0994$. Using the normal approximation to the binomial yields .0957, a difference of .0037.

8.16 (a) $\mu = 4.5$. (b) $\sigma = \sqrt{3.15} = 1.77482$. (c) If $p = 0.1$, then $\sigma = \sqrt{1.35} = 1.16190$. If $p = 0.01$, then $\sigma = \sqrt{0.1485} = 0.38536$. As p gets close to 0, σ gets closer to 0.

8.17 (a) $\mu = (20)(.8) = 16$. (b) $\sigma = \sqrt{(20)(.8)(.2)} = \sqrt{3.2} = 1.789$. (c) For $p = 0.99$, $\sigma = \sqrt{(20)(.99)(.01)} = \sqrt{0.198} = .445$. As the probability of success gets closer to 1, the standard deviation decreases. (Note that as p approaches 1, the probability histogram of the binomial distribution becomes increasingly skewed, and thus there is less and less chance of seeing an observation of the binomial at an appreciable distance from the mean.)

8.18 $\mu = 2.5$, $\sigma = \sqrt{1.875} = 1.36931$.

8.19 (a) $X =$ the number of people in the sample of 400 adult Richmonders who approve of the President's reaction. X is approximately binomial because the sample size is small compared to the population size (all adult Richmonders), and as a result, the individual responses may be considered independent and the probability of success (approval) remains essentially the same from trial to trial. $n = 400$ and $p = .92$.
(b) $P(X \le 358) = \mathtt{binomcdf}\ (400, \ .92, \ 358) = .0441$.
(c) $\mu = (400)(.92) = 368$, $\sigma = \sqrt{(400)(.92)(.08)} = \sqrt{29.44} = 5.426$.
(d) $P(X \le 358) \approx P(Z \le \frac{358 - 368}{\sqrt{29.44}}) = P(Z \le -1.843) \approx .0327$. The approximation is not very accurate (note that p is close to 1).

8.20 (a) Yes. Assuming that the sample size is small compared to the population size—a reasonable assumption if your area is heavily populated—this study satisfies the requirements of a binomial setting. $n = 200$ because 200 households are surveyed, and $p = 0.4$ because the probability of success (being committed to eating nutritious food away from home) is 40% and remains essentially the same from trial to trial.
(b) $\mu = (200)(.4) = 80$. $\sigma = \sqrt{200(.4)(.6)} = \sqrt{48} = 6.923$.
(c) $np = 80$ and $n(1 - p) = 120$, so the rule of thumb is satisfied here. $P(75 \le X \le 85) \approx P(\frac{75 - 80}{\sqrt{48}} \le Z \le \frac{85 - 80}{\sqrt{48}}) = P(-0.72 \le Z \le 0.72) = 0.5285$.

8.21 The command $\mathtt{cumSum}\ (\mathrm{L}_2) \to \mathrm{L}_3$ calculates and stores the values of $P(X \le x)$ for $x = 0, 1, 2, ..., 12$. The entries in L_3 and the entries in L_4 defined by $\mathtt{binomcdf}\ (12, \ .75, \ \mathrm{L}_1) \to \mathrm{L}_4$ are identical (see below).

L2	L3	L1	4
6E-8	6E-8	6E-8	
2.1E-6	2.2E-6	2.2E-6	
3.5E-5	3.8E-5	3.8E-5	
3.5E-4	3.9E-4	3.9E-4	
.00239	.00278	.00278	
.01147	.01425	.01425	
.04015	.0544	.0544	

L4 = (5.960464478...

L2	L3	L4	4
.10324	.15764	.15764	
.19358	.35122	.35122	
.2581	.60932	.60932	
.23229	.84162	.84162	
.12671	.96832	.96832	
.03168	1	1	
------	------		

L4(14) =

The first screen holds the probabilities for $X = 0, 1, 2, 3, 4, 5, 6$. The second holds the probabilities for $X = 7, 8, 9, 10, 11, 12$.

8.22 (a) We simulate 10 observations of $X =$ number of defective switches (for which $n = 10$, $p = .1$) by using the command $\mathtt{randBin}\ (1, \ .1, \ 10) \to \mathrm{L}_1 \colon$ sum $(\mathrm{L}_1) \to \mathrm{L}_2 \ (1)$. (Press ENTER 10 times.) The observations for one sample simulation are: 0, 0, 4, 0, 1, 0, 1, 0, 0, 1. For these data, $\bar{x} = .7$. To generate 25/50 observations of x, replace 10 with 25/50 in the $\mathtt{randBin}$ command above. As the number of observations increases, the resulting $\bar{x}$ should approximate the known mean $\mu = 1$ more closely, by the law of large numbers.

(b) Simulate 10 observations of X = number of free throws Corinne makes (where the number of trials is $n = 12$ and the probability of a basket on a given trial is $p = .75$) by using the command `randBin (1, .75, 12) → L₁: sum (L₁) → L₂ (1)`. (Press ENTER 10 times.) The observations for one sample simulation are: 9, 8, 10, 10, 9, 8, 10, 9, 8, 8. For these data, $\bar{x} = 8.9$. Compare this with the known mean $\mu = np = (12)(.75) = 9$. To generate 25/50 observations of $\bar{x}$, replace 10 with 25/50 in the `randBin` command above. As the number of observations increases, the resulting $\bar{x}$ should approximate the known mean $\mu = 1$ more closely, by the law of large numbers.

8.23 There are $n = 15$ people on the committee, and the probability that a randomly selected person is Hispanic is $p = .3$. Let 0, 1, 2 $\Leftrightarrow$ Hispanic and let 3–9 $\Leftrightarrow$ non-Hispanic. Use the random digit table. Or, using the calculator, repeat the command 30 times: `randBin (1, .3, 15) → L₁: sum (L₁) → L₂ (1)` where 0 = non-Hispanic, and 1 = Hispanic. Our frequencies were:

1	2	4	5	11	3	4
0	1	2	3	4	5	6	7	8

For this simulation, the relative frequency of 3 or fewer Hispanics was $7/30 = .233$. Compare this with the theoretical result: $P(X \leq 3) = 0.29687$, where X = number of Hispanics on the committee.

8.24 The probability of a success (having four or more credit cards) in this case is $p = 0.33$. Let the two-digit groups 00, 01, 02, ..., 32 represent students who have four or more credit cards, and let 33, 34, 35, ..., 99 represent students who do not. Choose 30 two-digit groups from the table of random digits (Table B). Starting at Line 141 in the table, for example, we get

96	76	73	59	64	23	82	29	60	12
94	59	16	51	94	50	84	25	33	72
72	82	95	02	32	97	89	26	34	08

The total number of "students with four or more credit cards" here is 9. Repeat this process a total of 30 times to obtain 30 observations of the binomial variable X (# of students with four or more cards) and estimate $P(X > 12)$ by the relative frequency of observations of X that exceed 12.

If the TI-83 is used, we can generate 30 observations of X by the command `randBin (1, .33, 30) → L₁: sum (L₁) → L₂ (1)`, with ENTER being pressed 30 times to produce 30 simulated observations. A sample simulation yielded the following results:

10	13	11	11	6	13	9	9	11	11
10	6	8	6	8	7	13	10	13	10
6	13	3	14	9	9	11	8	14	12

Here, $P(X > 12) = 7/30 = 0.233$. The actual value of $P(X > 12)$ is $1 -$ `binomcdf (30, .33, 12)` $= 0.1563$.

8.25 The sample size is $n = 10$, and the probability that a randomly selected employed woman has never been married is $p = 0.25$. Let 0 $\Leftrightarrow$ never married, let 1, 2, 3, $\Leftrightarrow$ married, and use Table B. Or, using the calculator, repeat the command `randBin (1, .25, 10) → L₁: sum (L₁)`. Our results for 30 repetitions were:

2	3	14	5	4	2
0	1	2	3	4	5

Our relative frequency of "2 or fewer never married" was $19/30 = .63$. The actual value of $P(X \le 2) = \texttt{binomcdf (10, .25, 2)} = .525$.

8.26 (a) The probability of drawing a white chip is $\frac{15}{50} = \frac{3}{10} = .3$. The number of white chips in 25 draws is $B(25, .3)$. Therefore, the expected number of white chips is $(25)(.3) = 7.5$.

(b) The probability of drawing a blue chip is $\frac{10}{50} = \frac{2}{10} = .2$. The number of blue chips in 25 draws is $B(25, .2)$. Therefore, the standard deviation of the number of blue chips is $\sqrt{(25)(.2)(.8)} = 2$.

(c) Let the digits 0, 1, 2, 3, 4 $\Leftrightarrow$ red chip, 5, 6, 7 $\Leftrightarrow$ white chip, and 8, 9 $\Leftrightarrow$ blue chip. Draw 25 random digits from Table B and record the number of times that you get chips of various colors. If you are using the TI-83, you can draw 25 random digits using the command `randInt (0, 9, 25) → L₁`. Repeat this process 30 times (or however many times you like) to simulate multiple draws of 25 chips. A sample simulation of a single 25-chip draw using the TI-83 yielded the following result:

Digit	0	1	2	3	4	5	6	7	8	9
Frequency	4	3	4	2	1	2	0	2	1	6

This corresponds to drawing 14 red chips, 4 white chips, and 7 blue chips.

(d) The expected number of blue chips is $(25)(.2) = 5$, and the standard deviation $= 2$ by part (b). It seems extremely likely that you will draw at most 9 blue chips. The actual probability is `binomcdf (25, .2, 9)` $= .9827$.

(e) It seems virtually certain that you will draw 15 or fewer blue chips; the probability is even larger than in part (d). The actual probability is `binomcdf (25, .2, 15)` $= .999998$.

8.27 The count of 0s among n random digits has a binomial distribution with $p = 0.1$. (a) P(at least one 0) $= 1 - P$(no 0) $= 1 - (0.9)^5 = 0.40951$. (b) $\mu = (40)(0.1) = 4$.

8.28 (a) $n = 20$ and $p = 0.25$. (b) $\mu = 5$. (c) $\binom{20}{5} (0.25)^5 (0.75)^{15} = 0.20233$.

8.29 (a) $n = 5$ and $p = 0.65$. (b) X takes values from 0 to 5. (c) $P(X = 0) = .0053$, $P(X = 1) = .0488$, $P(X = 2) = .1815$, $P(X = 3) = .3364$, $P(X = 4) = .3124$, $P(X = 5) = .11603$. Histogram below. (d) $\mu = (5)(.65) = 3.25$, $\sigma = \sqrt{(5)(.65)(.35)} = \sqrt{1.1375} = 1.067$.

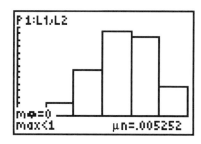

8.30 (a) The probability that all are assessed as truthful is $\binom{12}{0} (0.2)^0 (0.8)^{12} = 0.06872$; the probability that at least one is reported to be a liar is $1 - 0.06872 = 0.93128$.

(b) $\mu = 2.4$, $\sigma = \sqrt{1.92} = 1.38564$.

(c) $P(X < 2.4) = P(X \le 2) = \texttt{binomcdf(12, .2, 2)} = .5583$.

8.31 In this case, $\mu = (200)(.4) = 80$ and $\sigma = \sqrt{(200)(.4)(.6)} = 6.9282$. Using the normal approximation to the binomial distribution, $P(X \geq 100) \approx P(Z \geq 2.89) = .0019$. Using the exact binomial distribution, $P(X \geq 100) = 1 - P(X \leq 99) = 1 - \texttt{binomcdf}\ (200,\ .4,\ 99) = .0026$. Regardless of how we compute the probability, this is strong evidence that the local percentage of households concerned about nutrition is higher than 40%.

8.32 (a) There are 150 independent observations, each with probability of success (response) = .5. (b) $\mu = (150)(.5) = 75$ responses. (c) $P(X \leq 70) \approx P(Z \leq -0.82) = .2061$; using unrounded values and software yields .2071. (d) Use 200, since $(200)(.5) = 100$.

8.33 (a) Let X = the number of orders shipped on time. X is approximately $B(100, .9)$, since there are 100 independent observations, each with probability .9 (the population size is more than 10 times the sample size). $\mu = (100)(.9) = 90$ and $\sigma = \sqrt{(100)(.9)(.1)} = 3$. The normal approximation can be used, since the second rule of thumb is *just* satisfied; $n(1 - p) = 10$. $P(X \leq 86) \approx P(Z \leq -1.33) = .0918$; the software value is .0912.

(b) Even when the claim is correct, there will be some variation in the sample proportions. In particular, in about 9% of all samples, we can expect to see 86 or fewer orders shipped on time.

8.34 (a) X has a binomial distribution with $n = 20$ and $p = 0.99$.

(b) $P(X = 20) = 0.81791$; $P(X < 20) = 0.18209$.

(c) $\mu = 19.8$, $\sigma = \sqrt{0.198} = 0.44497$.

8.35 Identical to Exercise 8.7, except that in this case we calculate $P(X \leq 11)$ for the 30 simulated observations. The actual value of $P(X \leq 11)$ is $\texttt{binomcdf}\ (20,\ .8,\ 11) = .00998$ or approximately .01.

8.36 (a) $P(X = 3) = .0613$. (b) $P(X \leq 2) = .9257$. (c) $P(X < 2) = P(X \leq 1) = .7206$. (d) $P(3 \leq X \leq 5) = P(X \leq 5) - P(X \leq 2) = .9999 - .9257 = .0742$. (e) $P(X < 2\ \text{or}\ X > 5) = P(X < 2) + P(X > 5) = P(X \leq 1) + (1 - P(X \leq 5)) = .7206 + (1 - .9999) = .7207$.

8.37 (a) Geometric setting; success = tail, failure = head; trial = flip of coin; $p = 1/2$.

(b) Not a geometric setting. You are not counting the number of trials before the first success is obtained.

(c) Geometric setting; success = jack, failure = any other card; trial = drawing of a card; $p = 4/52 = 1/13$. (Trials are independent because the card is replaced each time.)

(d) Geometric setting, success = match all 6 numbers, failure = do not match all 6 numbers; trial = drawing on a particular day; the probability of success is the same for each trial; $p = \dfrac{6!}{\binom{44}{6}} = .000102$; and trials are independent because the setting of a drawing is always the same and the results on different drawings do not influence each other.

(e) Not a geometric setting. The trials (draws) are not independent because you are drawing without replacement. Also, you are interested in getting 3 successes, rather than just the first success.

8.38 (a) The four conditions of a geometric setting hold, with probability of success 5 1/2.

(b) and (d)

X	1	2	3	4	5	...
P(X)	.5	.25	.125	.0625	.03125	
c.d.f.	.5	.75	.875	.9375	.96875	

(c) and (d)

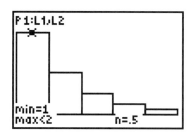 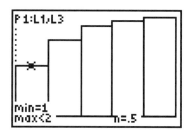

(e) Sum $= \dfrac{a}{1-r} = \dfrac{.5}{1-r} = 1.$

8.39 (a) X = number of drives tested in order to find the first defective. Success = defective drive. This is a geometric setting because the trials (tests) on successive drives are independent, $p = .03$ on each trial, and X is counting the number of trials required to achieve the first success.

(b) $P(X = 5) = (1 - .03)^{5-1}(.03)$

$\qquad = (0.97)^4(.03)$

$\qquad = .0266.$

(c)

X	1	2	3	4
$P(X)$	.03	.0291	.0282	.0274

8.40 Parts (a), (c), and (d) of Exercise 8.37 constituted geometric settings. (a) $P(X = 4) = (.5)^3(.5) = (.5)^4 = .0625$. (c) $P(X = 4) = (\frac{12}{13})^3(\frac{1}{13}) = .0605$. (d) $P(X = 4) = (.999898)^3(.000102) = .0001$.

8.41 (a) X = number of flips required in order to get the first head. X is a geometric random variable with $p = .5$.

(b) $P(X = x) = (.5)^{x-1}(.5) = (.5)^x$ for $x = 1, 2, 3, 4, \ldots$

x	1	2	3	4	5	$\ldots$
$P(X = x)$	.5	.25	.125	.0625	.03125	$\ldots$

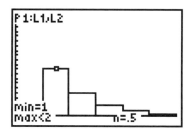

(c) $P(X \le x) = (.5)^1 + \ldots + (.5)^x$ for $x = 1, 2, 3, 4, \ldots$

x	1	2	3	4	5	$\ldots$
$P(X \le x)$	.5	.75	.875	.9375	.96875	$\ldots$

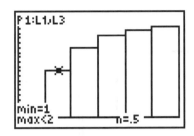

8.42 (a) $P(X > 10) = (1 - \frac{1}{12})^{10} = (\frac{11}{12})^{10} = .419$.

(b) $P(X > 10) = 1 - P(X \le 10) = 1 - \texttt{geometcdf(1/12, 10)} = .419$.

8.43 (a) The cumulative distribution histogram (out to $X = 10$) for rolling a die is shown below. Note that the cumulative function value for $X = 10$ is only .8385. Many more bars are needed for it to reach a height of 1.

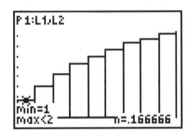 

(b) $P(X > 10) = (1 - 1/6)^{10} = (5/6)^{10} = 1615$. (c) The smallest positive integer k for which $P(X \le k) > .99$ is $k = 26$ (see second calculator screen above).

8.44 Let X = the number of applicants who need to be interviewed in order to find one who is fluent in Farsi. X is geometric with $p = 4\% = .04$. (a) $\mu = \frac{1}{p} = \frac{1}{.04} = 25$. (b) $P(X > 25) = (1 - .04)^{25} = (.96)^{25} = .3604$; $P(X > 40) = (.96)^{40} = .1954$.

8.45 (a) Assumptions needed for the geometric model to apply are that the shots are independent, and that the probability of success is the same for each shot. A "success" is a missed shot, so the probability of success is $p = 0.2$. The four conditions for a geometric setting are satisfied.

(b) The first "success" (miss) is the sixth shot, so $X = 6$ and $P(X = 6) = (1 - p)^{n-1} p = (.8)^5 (.2) = .0655$.

(c) $P(X \le 6) = 1 - P(X > 6) = 1 - (1 - p)^6 = 1 - (.8)^6 = .738$ or $P(X \le 6) = \texttt{geometcdf (.2, 6)} = .738$.

8.46 (a) Out of 8 possible outcomes, HHH and TTT do not produce winners. So P(no winner) = 2/8 = .25.

(b) P(winner) = 1 − .25 = .75.

(c) Let X = number of coin tosses until someone wins. Then X is geometric because all four conditions for a geometric setting are satisfied.

(d)

X	1	2	3	4	5	...
P(X)	.75	.1875	.04688	.01172	.00293	
c.d.f.	.75	.9375	.9844	.9961	.9990	

(e) $P(X \leq 2) = .9375$ from the table. (f) $P(X > 4) = (.25)^4 = .0039$. (g) $\mu = 1/p = 1/.75 = 1.33$.
(h) Let $1 \Leftrightarrow$ heads and $0 \Leftrightarrow$ tails, and enter the command `randInt (0, 1, 3)` and press
ENTER 25 times. In our simulation, we recorded the following frequencies:

X	1	2	3
Freq.	21	3	1
Rel. freq.	.84	.12	.04

These compare with the calculated probabilities of .75, .1875, and .04688, respectively. A larger number of trials should result in somewhat better agreement (Law of Large Numbers).

8.47 (a) Geometric setting; X = number of marbles you must draw to find the first red marble. We choose geometric in this case because the number of trials (draws) is the variable quantity.

(b) $P = 20/35 = 4/7$ in this case, so

$$P(X = 2) = (1 - 4/7)^{2-1}(4/7) = (3/7)(4/7) = 12/49 = .2449$$
$$P(X \leq 2) = 4/7 + (3/7)(4/7) = 4/7 + 12/49 = 40/49 = .8163$$
$$P(X > 2) = (1 - 4/7)^2 = (3/7)^2 = 9/49 = .1837$$

(c) Use the TI-83 commands `seq (X, X, 1, 20) → L₁`, `geometpdf (4/7, L₁) → L₂`, `cumSum (L₂) → L₃` (or `geometcdf (4/7, L₁) → L₃`).

X	1	2	3	4	5	6	7	8	9	10
P(X)	.571	.245	.105	.045	.019	.008	.004	.002	.001	.000
F(X) c.d.f.	.571	.816	.921	.966	.986	.994	.997	.999	.99951	.9998

X	11	12	13	14	15	16	17	18	19	20
P(X)	.000	.000	.000	.000	.000	.000	.000	.000	.000	.000
F(X)	.9999	.9999	.9999	1	1	1	1	1	1	1

(d) The probability distribution histogram is below left; the cumulative distribution is below right.

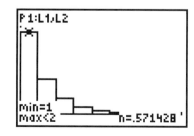
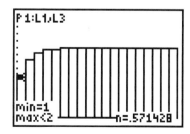

8.48 (a) No. Since the marbles are being drawn *without* replacement and the population (the set of all marbles in the jar) is so small, the results of any draw will clearly be dependent upon the results of previous draws. Also, the geometric variable measures the number of trials required to get the *first* success; here, we are looking for the number of trials required to get the *second* success.

(b) No. Even though the results of the draws are now independent, the variable being measured is still not the geometric variable.

(c) The probability of getting a red marble on any draw is $\frac{20}{35} = \frac{4}{7}$. Let the digits 0, 1, 2, 3 $\Leftrightarrow$ a red marble is drawn, 4, 5, 6 $\Leftrightarrow$ some other color marble is drawn, and 7, 8, 9 $\Leftrightarrow$ digit is disregarded. Start choosing random digits from Table B, or use the TI-83 command `randInt (0, 9, 1)` repeatedly. After two digits in the set 0, 1, 2, 3 have been chosen, stop the process and count the number of digits in the set 0, 1, 2, 3, 4, 5, 6 that have been chosen up to that point; this represents an observation of X. Repeat the process until the desired number of observations of X have been obtained. Here are some sample simulations using the TI-83 (with R = red marble, O = other color marble, D = disregard):

$$
\begin{array}{cccccc}
7 & 0 & 4 & 3 & & X = 3 \\
\mathbf{D} & \mathbf{R} & \mathbf{O} & \mathbf{R} & & \\
\\
9 & 0 & 8 & 6 & 2 & X = 3 \\
\mathbf{D} & \mathbf{R} & \mathbf{D} & \mathbf{O} & \mathbf{R} & \\
\\
9 & 7 & 3 & 2 & & X = 2 \qquad \text{etc.} \\
\mathbf{D} & \mathbf{D} & \mathbf{R} & \mathbf{R} & &
\end{array}
$$

8.49 (a) Success = getting a correct answer. X = number of questions Carla must answer in order to get the first correct answer. $p = 1/5 = .2$ (all 5 choices equally likely to be selected).

(b) $P(X = 5) = (1 - 1/5)^{5-1}(1/5) = (4/5)^4(1/5) = .082$.

(c) $P(X > 4) = (1 - 1/5)^4 = (4/5)^4 = .4096$.

(d)

X	1	2	3	4	5	. . . .
P(X)	.2	.16	.128	.1024	.082	. . . .

(e) $\mu = 1/(1/5) = 5$.

8.50 (a) If "success" = son and p (success) = .5, then the average number of children per family is $\mu = 1/p = 1/.5 = 2$.

(b) If the average size of the family is 2, and the last child is a boy, then the average number of girls per family is 1.

(c) Let even digit = boy, and odd digit = girl. Read random digits until an even digit occurs. Count number of digits read. Repeat many times, and average the counts. Beginning on line 101 in the random digit table and simulating 50 trials, the average number of children per family is 1.96, and the average number of girls is .96. These are very close to the expected values.

8.51 (a) Letting G = girl and B = boy, the outcomes are: {G, BG, BBG, BBBG, BBBB}. Success = having a girl.

(b) X = number of boys can take values of 0, 1, 2, 3, or 4. The probabilities are calculated by using the multiplication rule for independent events:

$$P(X = 0) = 1/2$$
$$P(X = 1) = (1/2)(1/2) = 1/4$$
$$P(X = 2) = (1/2)(1/2)(1/2) = 1/8$$
$$P(X = 3) = (1/2)(1/2)(1/2)(1/2) = 1/16$$
$$P(X = 4) = (1/2)(1/2)(1/2)(1/2) = 1/16$$

X	0	1	2	3	4
P(X)	1/2	1/4	1/8	1/16	1/16

Note that $\Sigma P(X) = 1$.

(c) Let Y = number of children produced until first girl is seen. Then Y is a geometric variable for $Y = 1$ up to $Y = 4$, but then "stops" because the couple plans to stop at 4 children if it does not see a girl by that time. By the multiplication rule,

$$P(Y = 1) = 1/2$$
$$P(Y = 2) = 1/4$$
$$P(Y = 3) = 1/8$$
$$P(Y = 4) = 1/16$$

Note that the event $\{Y = 4\}$ can only include the outcome BBBG. BBBB must be discarded. The probability distribution table would begin

Y	1	2	3	4
P(Y)	1/2	1/4	1/8	1/16

But note that this table is incomplete and this is not a valid probability model since $\Sigma P(Y) < 1$. The difficulty lies in the way Y was defined. It does not include the possible outcome BBBB.

(d) Let Z = number of children per family. Then

Z	1	2	3	4
P(Z)	1/2	1/4	1/8	1/16

and $\mu_z = \Sigma(Z \times P(Z)) = (1)(1/2) + (2)(1/4) + (3)(1/8) + (4)(1/8)$

$$= 1/2 + 1/2 + 3/8 + 1/2$$

$$= 1.875.$$

(e) $P(Z > 1.875) = P(2) + P(3) + P(4) = .5$.

(f) The only way in which a girl cannot be obtained is BBBB, which has probability 1/16. Thus the probability of having a girl, by the complement rule, is $1 - 1/16 = 15/16 = .938$.

8.52 Let $0 - 4 \Leftrightarrow$ girl and $5 - 9 \Leftrightarrow$ boy. Beginning with line 130 in the random digit table:

690	51	64	81	7871	74	0	951	784
B B G	B G	B G	B G	B B B G	B G	G	B B G	B B G
3	2	2	2	4	2	1	3	3

5 3	4	0	6 4	8 9 8 7 2	0	1	9 7 2	4	5 0	5 0
B G	G	G	B G	B B B B G	G	G	B B G	G	B G	B G
2	1	1	2	5	1	1	3	1	2	2

0	7 1	6 6 3	2	8 1
B	B G	B B G	G	B G
1	2	3	1	2

The average number of children is $52/25 = 2.08$. This compares with the expected value of 1.875.

8.53 We will approximate the expected number of children, μ, by making the mean $\bar{x}$ of 25 randomly generated observations of X. We create a suitable string of random digits (say of length 100) by using the command `randInt (0, 9, 100)` $\rightarrow L_1$. Now we scroll down the list L_1. Let the digits 0 to 4 represent a boy and 5 to 9 represent a girl. We read digits in the string until we get a "5 to 9" (girl) or until four "0 to 4"s (boys) are read, whichever comes first. In each case, we record X = the number of digits in the string = the number of children. We continue until 25 X-values have been recorded. Our sample string L_1 yielded the following values of X:

2 4 5 / 8 / 0 6 / 3 7 / 9 / 6 / 6 / 6 / 2 4 4 3 / 9 / 9 / 1 6 / 4 5 /
(3)   (1)   (2)   (2)   (1)   (1)   (1)   (1)   (4)   (1)   (1)   (2)   (2)

8 / 3 3 6 / 1 5 / 9 / 1 3 3 1 / 3 8 / 8 / 4 8 / 3 7 / 1 1 9 / 5 / 8 /
(1)   (3)   (2)   (1)   (4)   (2)   (1)   (2)   (2)   (3)   (1)   (1)

This yields $\bar{x} = 45/25 = 1.8$, compared with the known mean $\mu = 1.875$.

8.54 (a) Recall that if p = probability of success, then $\mu = 1/p$. Then the table is as follows:

X	.1	.2	.3	.4	.5	.6	.7	.8	.9
Y	10	5	3.33	2.5	2	1.67	1.43	1.25	1.1

(b)

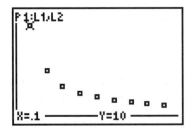

(c) Assuming the power function model $y = ab^x$, we transform the data by taking logs of both sides: $\log y = a + b \log x$. We thus compute and store the logs of the x's and y's.

(d) Here is the plot of the transformed data log y vs. log x:

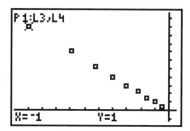

(e) The correlation is $r = -.999989$, or approximately -1.

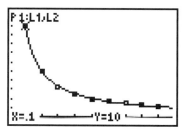

(f) The equation of the power function is $\hat{y} = 1x^{-1} = 1/x$.

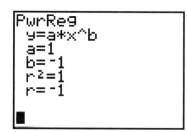

(g) The power function illustrates the fact that the mean of a geometric random variable is the *reciprocal* of the probability p of success: $\mu = 1/p$.

8.55 (a) No. It is not reasonable to assume that the opinions of a husband and wife (especially on such an issue as mothers working outside the home) are independent.

(b) No. The sample size (25) is so small compared to the population size (75) that the probability of success ("Yes") will substantially change as we move from person to person within the sample. The population should be at least 10 times larger than the sample in order for the binomial setting to be valid.

8.56 $P(\text{alcohol related fatality}) = 346/869 = .398$. If $X =$ number of alcohol related fatalities, then X is $B(25, .398)$. $\mu = np = 25(.398) = 9.95$. $\sigma = \sqrt{(9.95)(.602)} = 2.45$. $P(X \le 5) =$ binomcdf(25, .398, 5) $= .0307$.

8.57 (a) The distribution of $X = B(7, .5)$ is symmetric; the shape depends on the value of the probability of success. Since .5 is halfway between 0 and 1, the histogram is symmetric.

(b) With the values of $X = 0, 1, \ldots, 7$ in L_1, define L_2 to be `binompdf (7, .5)`. Then the probability table for B(7, .5) is installed in L_1 and L_2. Here is a histogram of the p.d.f.:

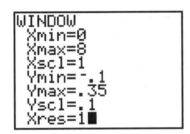 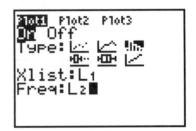

Now, define L_3 to be `binomcdf (7, .5, L₁)`. Then the cdf for the B(7, .5) distribution is installed in L_3. Here is the histogram of the cdf:

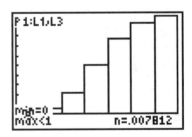

(c) $P(X = 7) = $ `binompdf (7, .5, 7)` $= .0078125$.

8.58 (a) In our simulation, we obtained the following results:

Outcome	1	2	3	4	5	6	7	8
Frequency	28	17	2	2				1
Rel. freq.	.56	.34	.04	.04				.02

(b) We observed heads on the first toss 56% of the time. Our estimate of the probability of heads on the first toss is 0.5.

(c) An estimate of the probability that the first head appears on an odd-numbered toss is 2/3.

8.59 (a) $p = .2$.

(b)

Result	Probability
SSSS	$(.2)(.2)(.2)(.2) = .0016$
SSSF	$(.2)(.2)(.2)(.8) = .0064$
SSFS	$(.2)(.2)(.8)(.2) = .0064$
SFSS	$(.2)(.8)(.2)(.2) = .0064$
FSSS	$(.8)(.2)(.2)(.2) = .0064$
SSFF	$(.2)(.2)(.8)(.8) = .0256$
SFSF	$(.2)(.8)(.2)(.8) = .0256$
SFFS	$(.2)(.8)(.8)(.2) = .0256$
FSFS	$(.8)(.2)(.8)(.2) = .0256$
FSSF	$(.8)(.2)(.2)(.8) = .0256$
FFSS	$(.8)(.8)(.2)(.2) = .0256$
SFFF	$(.2)(.8)(.8)(.8) = .1024$
FSFF	$(.8)(.2)(.8)(.8) = .1024$
FFSF	$(.8)(.8)(.2)(.8) = .1024$
FFFS	$(.8)(.8)(.8)(.2) = .1024$
FFFF	$(.8)(.8)(.8)(.8) = .4096$

(c)

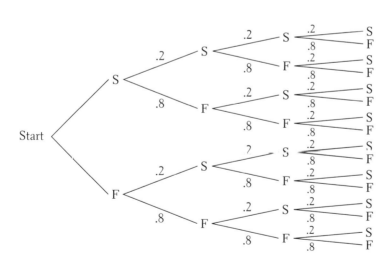

(d) SSFF, SFSF, SFFS, FSFS, FSSF, FFSS.

(e) Each outcome has probability .0256. The probabilities all involve the same number of each factor; they are simply multiplied in different orders, which will not affect the product.

8.60 Let X = the number of schools out of 20 that say they have a soft drink contract. X is binomial with $n = 20$ and $p = .62$.

(a) $P(X = 8) = \texttt{binompdf (20, .62, 8)} = .0249$.

(b) $P(X \le 8) = $ binomcdf (20, .62, 8) $ = .0381$.

(c) $P(X \ge 4) = 1 - P(X \le 3) = 1 - $ binomcdf (20, .62, 3) $ = .99998$.

(d) $P(4 \le X \le 12) = P(X \le 12) - P(X \le 3) = $ binomcdf (20, .62, 12) $-$ binomcdf (20, .62, 3) $ = .5108 - .0381 = .4727$.

(e) $X = $ the number of schools out of 20 that say they have a soft drink contract. Create the probability distribution table for X using the TI-83 by entering the values 0, 1, 2, ..., 20 into column L_1 and then entering the command binompdf (20, .62, L_1) $\to L_2$ to store the probabilities in L_2. The pdf table is displayed in the three screens below.

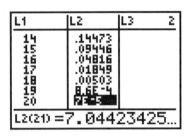

(f) The TI-83 version of the histogram is given below. (Note that some of the bars are too short to appear in the histogram.)

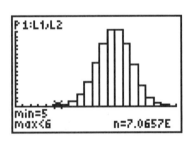

8.61 Let $X = $ the number of southerners out of 20 that believe they have been healed by prayer. X is binomial with $n = 20$ and $p = .46$.

(a) $P(X - 10) = $ binompdf (20, .46, 10) $ = .1652$.

(b) $P(10 < X < 15) = P(11 \le X \le 14) = P(X \le 14) - P(X \le 10) = $ binomcdf (20, .46, 14) $-$ binomcdf (20, .46, 10) $ = .9917 - .7209 = .2708$.

(c) $P(X > 15) = 1 - P(X \le 15) = 1 - $ binomcdf (20, .46, 15) $ = 1 - .9980 = .0020$.

(d) $P(X < 8) = P(X \le 7) = $ binomcdf (20, .46, 7) $ = .2241$.

8.62 (a) $P(X = 2) = .2753$.

(b) $P(X \ge 2) = P(X = 2) + \ldots + P(X = 7) = .4234$.

(c) $P(X < 2) = P(X = 0) + P(X = 1) = .5767$.

(d) $P(2 \le X \le 5) = P(X = 2) + \ldots + P(X = 5) = .4230$.

8.63 X is geometric with $p = .325$. $1 - p = .675$. (a) $P(X = 1) = .325$. (b) $P(X \le 3) = .325 + (.675)$ $(.325) + (.675)^2 (.325) = .69245$. (c) $P(X > 4) = (.675)^4 = .208$. (d) The expected number of at-bats until Roberto gets his first hit is $\mu = 1/p = 1/.325 = 3.08$. (e) To do this, use the commands seq (x, x, 1, 10) $\to L_1$, geometpdf (.325, L_1) $\to L_2$, and geometcdf (.325, L_1) $\to L_3$. (f) See the top of the facing page for the plot of the histogram.

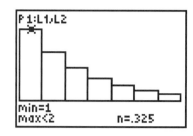 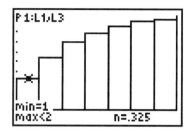

8.64 (a) By the $68-95-99.7$ rule, the probability of any one observation falling within the interval $\mu - \sigma$ to $\mu + \sigma$ is .68. Let X = the number of observations out of 5 that fall within this interval. Assuming that the observations are independent, X is $B(5, .68)$. Then, $P(X = 4) =$ binompdf (5, .68, 4) = .3421

(b) By the $68-95-99.7$ rule, 95% of all observations fall within the interval $\mu - 2\sigma$ to $\mu + 2\sigma$. Thus, 2.5% (half of 5%) of all observations will fall above $\mu + 2\sigma$. Let X = the number of observations that must be taken before we observe one falling above $\mu + 2\sigma$. Then X is geometric with $p = .025$. $P(X = 4) = (1 - .025)^3(.025) = (.975)^3(.025) = .0232$.

8.65 $P(X \geq 1) = 1 - P(X = 0) = 1 - \binom{n}{0}p^0 (1 - p)^{n-0} = 1 - (1)(1)(1 - p)^n = 1 - (1 - p)^n$.

9

Sampling Distributions

9.1 $\mu = 2.5003$ is a parameter; $\bar{x} = 2.5009$ is a statistic.

9.2 $\hat{p} = 7.2\%$ is a statistic.

9.3 $\hat{p} = 48\%$ is a statistic; $p = 52\%$ is a parameter.

9.4 Both $\bar{x}_1 = 335$ and $\bar{x}_2 = 289$ are statistics.

9.5 (a) Since the proportion of times the toast will land butter-side down is 0.5, the result of 20 coin flips will simulate the outcomes of 20 pieces of falling toast (landing butter-side up or butter-side down).

(b) Answers will vary.

(c) Answers will vary; however, it is more likely that the center of this distribution will be close to 0.5, and it is more likely that the shape will be close to normal.

(d) Answers will vary.

(e) We obtain a more accurate representation of a sampling distribution when many samples are taken.

9.6 (a)

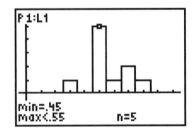

The results appear to be quite variable.

(b)

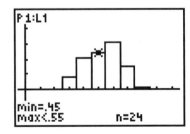

The center is close to 0.5, and the shape is approximately normal.

(c)

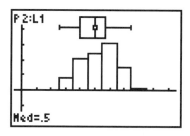

The median and mean are extremely close.

(d) The spread of the distribution did not seem to change. To decrease the spread, I would increase the number of trials, n. For example, use `randBin (50, .5)`.

9.7 (a) The scores will vary depending on the starting row. Note that the smallest possible mean is 61.75 (from the sample 58, 62, 62, 65) and the largest is 77.25 (from 73, 74, 80, 82).

(b)–(c) Answers will vary; shown below are two views of the sampling distribution. The first shows all possible values of the experiment (so the first rectangle is for 61.75, the next is for 62.00, etc.); the other shows values grouped from 61 to 61.75, 62 to 62.75, etc. (which makes the histogram less bumpy). The tallest rectangle in the first picture is 8 units; in the second, the tallest is 28 units.

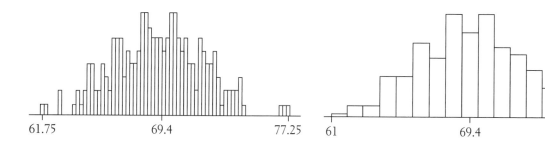

(d) There are $(10 \times 9)/2 = 45$ possible samples of size 2 that can be drawn from the population.

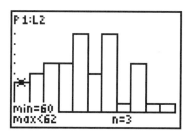

(e) The shapes and centers for the two distributions are roughly the same. However, the spread is a little larger for the distribution corresponding to $n = 2$. This distribution is somewhat more irregular, reflecting the fact that sample means based on samples of size 2 tend to be more variable than those based on samples of size 4.

9.8 (a) Table is on the next page; histogram not shown. (b) The histogram actually does *not* appear to have a normal shape. The sampling distribution is quite normal in appearance, but even a sample of size 100 does not *necessarily* show it. (c) The mean of $\hat{p}$ is 0.0981. The bias seems to be small. (d) The mean of the sampling distribution should be $p = 0.10$. (e) The mean would still be 0.10, but the spread would be smaller.

p	$\hat{p}$	Count	p	$\hat{p}$	Count	p	$\hat{p}$	Count
9	0.045	1	18	0.090	12	24	0.120	10
13	0.065	3	19	0.095	9	25	0.125	4
14	0.070	2	20	0.100	7	26	0.130	1
15	0.075	5	21	0.105	5	27	0.135	2
16	0.080	11	22	0.110	6	28	0.140	2
17	0.085	12	23	0.115	7	30	0.150	1

9.9 (a) Below, left. (b) For the 72 survival times, $\mu = 141.847$ days. (c) Means will vary with samples. (d) It would be unlikely (though not impossible) for all five $\bar{x}$ values to fall on the same side of μ. This is one implication of the unbiasedness of $\bar{x}$: Some values will be higher and some lower than μ. (But note, it is not necessarily half and half.) (e) Shown (below, right) is a stemplot for one set of 100 sample means, which approximates the sampling distribution of $\bar{x}$. This set of means varied from 85 to 225 days and had mean 138.1 and standard deviation 25.9 days. The mean of the (theoretical) sampling distribution would be μ. (f) Answers will vary.

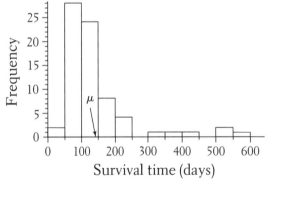

```
 8 | 5
 9 | 1789
10 | 1133589999
11 | 0000111367789
12 | 245566789
13 | 01112233666689999
14 | 00001111223689
15 | 001223355567889
16 | 02348
17 | 234567
18 | 14
19 | 223
20 |
21 |
22 | 5
```

9.10 (a) Large bias and large variability. (b) Small bias and small variability. (c) Small bias, large variability. (d) Large bias, small variability.

9.11 (a) Since the smallest number of total tax returns (i.e., the smallest population) is still more than 100 times the sample size, the variability will be (approximately) the same for all states.

(b) Yes, it will change—the sample taken from Wyoming will be about the same size, but the sample in, e.g., California will be considerably larger, and therefore the variability will decrease.

9.12 $\bar{x} = 64.5$ is a statistic; $\mu = 63$ is a parameter.

9.13 $\hat{p} = 4.5\% = .045$ is a statistic.

9.14 (a) Use digits 0 and 1 (or any other 2 of the 10 digits) to represent the presence of egg masses. Reading the first 10 digits from line 116, for example, gives YNNNN NNYNN—2 square yards with egg masses, 8 without—so $\hat{p} = 0.2$.

(b) The stemplot *might* look like the one below (which is close to the sampling distribution of $\hat{p}$).

(c) The mean would be $p = 0.2$. (d) 0.4.

```
0.0 | 00
0.0 | 55555
0.1 | 000000
0.1 | 5555
0.2 | 00
0.2 | 5
```

9.15 (a) $\mu = 3.5$, $\sigma = 1.708$.

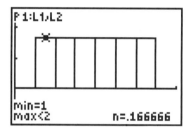

(b) This is equivalent to rolling a pair of fair, six-sided dice.

(c)

SRS of size 2	$\bar{x}$
1, 1	1
2, 1 1, 2	1.5
3, 1 2, 2 1, 3	2
4, 1 2, 3 3, 2 1, 4	2.5
5, 1 2, 4 3, 3 4, 2 1, 5	3
6, 1 2, 5 3, 4 4, 3 5, 2 1, 6	3.5
6, 2 3, 5 4, 4 5, 3 2, 6	4
6, 3 4, 5 5, 4 3, 6	4.5
6, 4 5, 5 4, 6	5
6, 5 5, 6	5.5
6, 6	6

(d) Histogram below. The center is identical to that of the population distribution. The shape is normal (symmetric and bell-shaped) rather than uniform. The spread is smaller than that of the population distribution; the probability of observing a value at some distance from the center remains constant for the population distribution but decreases with increasing distance for the histogram corresponding to $n = 2$.

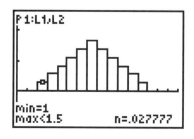

9.16 Answers will vary. A sample histogram is shown below. While the center of the distribution remains the same, the spread is smaller than that of the histogram in Exercise 9.15.

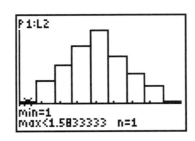

9.17 Assuming that the poll's sample size was less than 780,000–10% of the population of New Jersey—the variability would be practically the same for either population. (The sample size for this poll would have been considerably less than 780,000.)

9.18 (a) The digits 1 to 41 are assigned to adults who say that they have watched *Survivor II*. The program outputs a proportion of "Yes" answers. For (b), (c), (d), and (e), answers will vary; however, as the sample size increases from 5 to 25 to 100, the variability of the sample proportions should decrease.

9.19 (a) $\mu = p = 0.7$; $\sigma = \sqrt{\frac{p(1-p)}{n}} = \sqrt{\frac{(0.7)(0.3)}{1012}} = 0.0144$.

   (b) The population (all U.S. adults) is clearly at least 10 times as large as the sample (the 1012 surveyed adults).

   (c) $np = (1012)(.7) = 708.4 \geq 10$; $n(1-p) = (1012)(.3) = 303.6 \geq 10$.

   (d) $P(\hat{p} \leq .67) = P(Z \leq -0.25) = 0.0186$—this is a fairly unusual result if 70% of the population actually drinks the cereal milk.

   (e) Multiply the sample size by 4; we would need to sample $(1012)(4) = 4048$ adults.

   (f) It would probably be higher, since teenagers (and children in general) have a greater tendency to drink the cereal milk.

9.20 (a) $\mu = p = 0.4$, $\sigma = \sqrt{(0.4)(0.6) \div 1785} = 0.0116$. (b) The population (U.S. adults) is considerably larger than 10 times the sample size. (c) $np = 714$, $n(1-p) = 1071$—both are much bigger than 10. (d) $P(0.37 < \hat{p} < 0.43) = P(-2.586 < Z < 2.586) = 0.9904$. Over 99% of all samples should give $\hat{p}$ within $\pm 3\%$ of the true population proportion.

9.21 For $n = 300$: $\sigma = 0.02828$ and $P = 0.7108$. For $n = 1200$: $\sigma = 0.01414$ and $P = 0.9660$. For $n = 4800$: $\sigma = 0.00707$ and $P = 1$ (approximately). Larger sample sizes give more accurate results (the sample proportions are more likely to be close to the true proportion).

9.22 (a) The distribution is approximately normal with mean $\mu = p = 0.14$ and standard deviation $\sigma = \sqrt{\frac{p(1-p)}{n}} = \sqrt{\frac{(0.14)(0.86)}{500}} = 0.0155$. (b) 20% or more Harley owners is unlikely; $P(\hat{p} > 0.20) \approx P(Z > 3.87) < 0.0002$. There is a fairly good chance of finding at least 15% Harley owners; $P(\hat{p} > 0.15) \approx P(Z > 0.64) = 0.2611$.

9.23 (a) 0.86 (86%). (b) We use the normal approximation (Rule of Thumb 2 is *just* satisfied—$n(1-p) = 10$). The standard deviation is 0.03, and $P(\hat{p} \leq 0.86) = P(Z \leq -1.33) = 0.0918$. (*Note:* The exact probability is 0.1239.) (c) Even when the claim is correct, there will be some variation in sample proportions. In particular, in about 10% of samples we can expect to observe 86 or fewer orders shipped on time.

9.24 The calculation for Exercise 9.22 should be more accurate. This calculation is based on a larger sample size (500, as opposed to the 100 of Exercise 9.23). Rule of Thumb 2 is easily satisfied in Exercise 9.22 but just barely satisfied in Exercise 9.23.

9.25 (a) $\mu = p = 0.15$, $\sigma = \sqrt{(0.15)(0.85) \div 1540} = 0.0091$. (b) The population (U.S. adults) is considerably larger than 10 times the sample size (1540). (c) $np = 231$, $n(1 - p) = 1309$—both are much bigger than 10. (d) $P(0.13 < \hat{p} < 0.17) = P(-2.198 < Z < 2.198) = 0.9722$. (e) To achieve $\sigma = .403$, we need a sample *nine* times as large—about 13,860.

9.26 For $n = 200$: $\sigma = 0.02525$, and the probability is $P = 0.5704$. For $n = 800$: $\sigma = 0.01262$ and $P = 0.8858$. For $n = 3200$: $\sigma = 0.0631$ and $P = 0.9984$. Larger sample sizes give more accurate results (the sample proportions are more likely to be close to the true proportion).

9.27 $P(\hat{p} \geq .608) = P\left(Z \geq \dfrac{.608 - .6}{\sqrt{\frac{(.6)(.4)}{2500}}} \right) = P(Z \geq .8165)$

$$= .2071. \text{ (The exact answer is .213.)}$$

9.28 (a) $\mu = 0.52$, $\sigma = 0.02234$. (b) np and $n(1 - p)$ are 260 and 240 respectively. $P(\hat{p} \geq 0.50) = P(Z \geq -0.8951) = 0.8159$.

9.29 (a) $P(\hat{p} \leq 0.70) = P(Z \leq -1.155) = 0.1241$. (b) $P(\hat{p} \leq 0.70) = P(Z \leq -1.826) = 0.0339$. (c) The test must contain 400 questions. (d) The answer is the same for Laura.

9.30 (a) $np = (15)(0.3) = 4.5$—this fails Rule of Thumb 2. (b) The population size (316) is not at least 10 times as large as the sample size (50)—this fails Rule of Thumb 1. (c) $P(X \leq 3) =$ `binomcdf(15, .3, 3)` $= .2969$.

9.31 (a) $\mu = -3.5\%$, $\sigma = 26\%/\sqrt{5} = 11.628\%$. (b) $P(X \geq 5\%) \approx P(Z \geq 0.3269) = .3719$. (c) $P(\bar{x} \geq 5\%) \approx P(Z \geq 0.73) = 0.2327$. (d) $P(\bar{x} < 0) \approx P(Z < 0.30) = 0.6179$. Approximately 62% of all five-stock portfolios lost money.

9.32 (a) $P(X \geq 21) = P(Z \geq 0.4068) = 0.3421$. (b) $\mu = 18.6$, $\sigma = 5.9/\sqrt{50} = 0.8344$. This result is independent of distribution shape. (c) $P(\bar{x} \geq 21) \approx P(Z \geq 2.8764) = 0.0020$.

9.33 (a) $\sigma/\sqrt{3} \doteq 5.7735$ mg. (b) Solve $\sigma/\sqrt{n} = 3$: $\sqrt{n} = \frac{10}{3}$, so $n = 11.1$ or 12. The average of several measurements is more likely than a single measurement to be close to the mean.

9.34 (a) If we choose many samples, the average of the $\bar{x}$-values from these samples will be close to μ. (I.e., $\bar{x}$ is "correct on the average" in many samples.)

(b) The larger sample will give more information, and therefore more precise results; that is, $\bar{x}$ is more likely to be close to the population truth. Also, $\bar{x}$ for a larger sample is less affected by outliers.

9.35 $\bar{x}$ has approximately a N(1.6, 0.0849) distribution; the probability is $P(Z > 4.71)$—essentially 0.

9.36 $\bar{x}$ (the mean return) has approximately a N(9%, 4.174%) distribution; $P(\bar{x} > 15\%) = P(Z > 1.437) = 0.9247$; $P(\bar{x} < 5\%) = P(Z < -0.9583) = 0.1690$.

9.37 (a) N(123, 0.04619). (b) $P(Z > 21.65)$—essentially 0.

9.38 (a) Mean: 40.125, standard deviation: 0.001; normality is not needed. (b) No: We cannot complete $p(\bar{x} > 40.127)$ based on a sample of size 4 because the sample size must be *larger* to justify use of the central limit theorem if the distribution type is unknown.

9.39 (a) $P(X < 295) = P(Z < -1) = 0.8413$. (b) $P(\bar{x} < 295) = P(Z < -2.4495) = 0.0072$.

9.40 (a) N(55000, 4500/$\sqrt{8}$) = N(55000, 1591). (b) $P(Z < -2.011) = 0.0222$.

9.41 (a) N(2.2, 0.1941). (b) $P(\bar{x} < 2) \approx P(Z < -1.0304) = 0.1515$. (c) $P(\bar{x} < \frac{100}{52}) \approx P(Z < -1.4267) = 0.0768$.

9.42 $\mu - 1.645 \, \sigma/\sqrt{n} = 12.513$.

9.43 (a) $p = 68\% = .68$ is a parameter; $\hat{p} = 73\% = 0.73$ is a statistic. (b) $\mu = p = 0.68$, $\sigma = \sqrt{\frac{p(1-p)}{n}} = \sqrt{\frac{0.68(0.32)}{150}} = 0.0381$. (c) $P(\hat{p} \geq .73) \approx P(Z \geq 1.3128) = 0.0946$. There is a 10% (one in ten) chance that an observation of $\hat{p}$ greater than or equal to the observed value of .73 will be seen.

9.44 (a), (b) Answers will vary. In one simulation, we obtained a total of 9 simulated values of $\hat{p}$ less than or equal to 0.65, a percentage of 18%. (c) $\hat{p}$ has an approximately normal distribution. $\mu = p = 0.7$, $\sigma = \sqrt{\frac{p(1-p)}{n}} = \sqrt{\frac{0.7(0.3)}{100}} = 0.0458$. (d) $P(\hat{p} \leq 0.65) \approx P(Z \leq -1.091) = 0.1376$. This result is reasonably close to the 18% obtained in our simulation. (e) For $n = 1000$, $\hat{p}$ is again approximately normal, with $\mu = p = 0.7$, $\sigma = \sqrt{\frac{p(1-p)}{n}} = \sqrt{\frac{0.7(0.3)}{1000}} = 0.0145$. $P(\hat{p} \leq 0.65) \approx P(Z \leq -3.45) = 0.0003$. $\hat{p}$ is less variable for larger sample sizes, so the probability of seeing a value of $\hat{p}$ less than or equal to 0.65 decreases.

9.45 $2P(\bar{x} > 27.4) = 2P\left(Z > \frac{27.4 - 25}{7/\sqrt{10}}\right) \approx 2P(Z > 1.084) = 2(0.1392) = .2784$.

9.46 (a) $\hat{p}$ has an approximately normal distribution with $\mu = p = 0.47$, $\sigma = \sqrt{\frac{p(1-p)}{n}} = \sqrt{\frac{0.47(0.53)}{1025}} = 0.0156$.

(b) The middle 95% of all sample results will fall within $2\sigma \approx 0.0312$ of the mean 0.47, that is, in the interval 0.4388 to 0.5012.

(c) $P(\hat{p} < 0.45) \approx P(Z < -1.283) = 0.0998$.

9.47 The mean loss from fire, by definition, is the long-term average of *many* observations of the random variable X = fire loss. The behavior of X is much less predictable if only a small number of observations are made. If only 12 policies were sold, then the company would have no protection against the large expense that would be incurred if one of the 12 policyholders happened to lose his or her home. If thousands of policies were sold, then the average fire loss for these policies would be far more likely to be close to μ, and the company's profit would not be endangered by the few large fire-loss payments that it would have to make.

9.48 $P(\bar{x} > 260) = P\left(Z > \frac{260 - 250}{300/\sqrt{10000}}\right) \approx P(Z > 3.33) = 0.0004$.

9.49 (a) $P\left(Z > \frac{105 - 100}{15}\right) = P(Z > \frac{1}{3}) = 0.36944$. (b) Mean: 100; standard deviation: 1.93649. (c) $P\left(Z > \frac{105 - 100}{1.93649}\right) = P(Z > 2.5820) = 0.00491$. (d) The answer to (a) could be quite different; (b) would be the same (it does not depend on normality at all). The answer we gave for (c) would still be fairly reliable because of the central limit theorem.

9.50 (a) No—a count assumes only whole-number values, so it cannot be normally distributed. (b) N(1.5, 0.02835). (c) $P\left(\bar{x} > \frac{1075}{700}\right) = P(Z > 1.2599) = 0.10386$.

9.51 $\mu + 2.33\sigma/\sqrt{n} = 1.4625$.

9.52 (a) $np = (25000)(0.141) = 3525$.

(b) $P(X \geq 3500) = P\left(Z \geq \frac{3500 - 3525}{55.027}\right) = P(Z \geq -0.4543) = 0.6752$.

9.53 $P\left(\frac{750}{12} < \bar{x} < \frac{825}{12}\right) = P(-1.732 < Z < 2.598) = 0.95368$.

10

Introduction to Inference

10.1 (a) 44% to 50%.

(b) We do not have information about the whole population; we only know about a small sample. We expect our sample to give us a good estimate of the population value, but it will not be exactly correct.

(c) The procedure used gives an estimate within 3 percentage points of the true value in 95% of all samples.

10.2 (a) The sampling distribution of $\bar{x}$ is normal with mean $\mu = 280$ and standard deviation $\sigma/\sqrt{n} = 60/\sqrt{840} \approx 2.1$. (b) Below. (c) 2 standard deviations—$m \approx 4.2$. (d) Below; the confidence intervals drawn may vary, of course. (e) 95% (by the 68-95-99.7 rule).

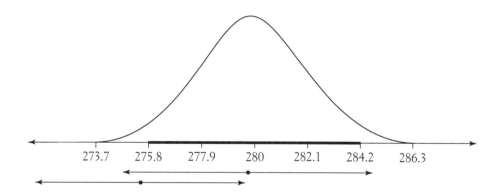

10.3 This is a statement about the *mean* score for all young men, not about individual scores. We are attempting only to estimate the center of the population distribution; the scores for individuals are much more variable. Also, "95%" is not a probability or a proportion; it is a confidence level.

10.4 (a) The sampling distribution of $\bar{x}$ is normal with mean μ and standard deviation $\sigma/\sqrt{n} = 0.4/\sqrt{50} = 0.05657$.

(b) See the sketch on the next page. For this problem, the "numbers" below the axis would be $\mu - 0.16971$, $\mu - 0.11314$, $\mu - 0.05657$, μ, etc.

(c) $m = 0.11314$ (2 standard deviations).

(d) 95%.

(e) See the next page. The actual confidence intervals drawn may vary.

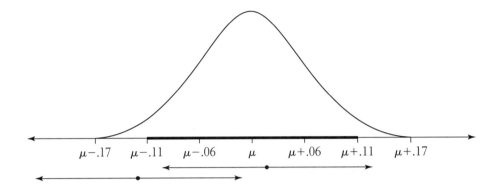

10.5 .84 ± 0.10, or .830 to .851 grams per liter.

10.6 11.78 ± 0.77, or 11.01 to 12.55 years.

10.7 (a) The distribution is slightly skewed to the right. (b) 224.002 ± 0.029, or 223.973 to 224.031.

2239	01
2239	66788889
2240	01
2240	589
2241	2

10.8 $\bar{x}$ = 123.8 bu/acre, and $\sigma_{\bar{x}}$ = $10/\sqrt{15}$ ≐ 2.582 bu/acre. (a)–(c) See the table below; the intervals are $\bar{x} \pm z^* \sigma_{\bar{x}}$. (d) The margin of error increases with the confidence level.

Conf. Level	z^*	Interval
90%	1.645	119.6 to 128.0 bu/acre
95%	1.960	118.7 to 128.9 bu/acre
99%	2.576	117.1 to 130.5 bu/acre

10.9 With n = 60, $\sigma_{\bar{x}}$ = $10/\sqrt{60}$ ≐ 1.291 bu/acre. (a) 95% confidence interval: $\bar{x} \pm 1.960\,\sigma_{\bar{x}}$ = 121.3 to 126.3 bu/acre. (b) Smaller: with a larger sample comes more information, which in turns gives less uncertainty ("noise") about the value of μ. (c) They will be narrower.

10.10 (a) 294 to 318.6. (b) 274.7 to 337.9. (c) Below—increasing confidence makes the interval wider.

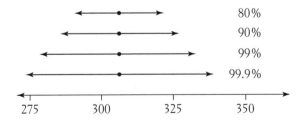

10.11 7.91—which is half the margin of error with n = 20.

10.12 (a) 10.00209 to 10.00251. (b) 22 (21.64).

10.13 68 (67.95).

10.14 35 (34.57).

10.15 (a) The computations are correct. (b) Since the numbers are based on a voluntary response, rather than an SRS, the methods of this section cannot be used—the interval does not apply to the whole population.

10.16 (a) The interval was based on a method that gives correct results 95% of the time.

(b) Since the margin of error was 2%, the true value of p could be as low as 49%. The confidence interval thus contains some values of p, which give the election to Bush.

(c) The proportion of voters that favor Gore is not random—either a majority favors Gore, or they don't. Discussing probabilities about this proportion has little meaning: the "probability" the politician asked about is either 1 or 0 (respectively).

10.17 (a) We can be 99% confident that between 63% and 69% of all adults favor such an amendment. We estimate the standard deviation of the distribution of $\hat{p}$ to be about $\sqrt{(0.66)(0.34)/1664} = 0.01161$; dividing 0.03 (the margin of error) by this gives $z^* = 2.58$, the critical value for a 99% confidence interval.

(b) The survey excludes people without telephones (a large percentage of whom would be poor), so this group would be underrepresented. Also, Alaska and Hawaii are not included in the sample.

10.18 No: The interval refers to the mean math score, not to individual scores, which will be much more variable (indeed, if more than 95% of students score below 470, they are not doing very well).

10.19 If we chose many samples of size 1548, then in about 95% of those samples, the percentage found would be within ± 3 percentage points of the true population percentage. That is, we are using a procedure that gives results within $\pm 3\%$ of the true percentage about 95% of the time.

10.20 (a) The intended population is hotel managers (perhaps specifically managers of hotels of the particular size range mentioned). However, because the sample came entirely from Chicago and Detroit, it may not do a good job of representing that larger population. There is also the problem of voluntary response.

(b) The central limit theorem allows us to say that $\bar{x}$ is approximately normal for a sample of this size ($n = 135$) no matter what the parent distribution is.

(c) 5.101 to 5.691.

(d) 4.010 to 4.876.

10.21 The sample size for women was more than twice as large as that for men. Larger sample sizes lead to smaller margins of error (with the same confidence level).

10.22 (a) The stemplot (see next page) shows no marked deviations from normality.

(b) $\bar{x} \doteq 25.67$ μm/hr, so the 90% confidence interval is $25.67 \pm 3.10 = 22.57$ to 28.77 μm/hr.

(c) Her interval is wider: To be more confident that our interval includes the true population parameter, we must allow a larger margin of error. So the margin of error for 95% confidence is larger than for 90% confidence.

```
1 | 124
1 | 8
2 | 2233
2 | 6789
3 | 034
3 | 55
4 | 0
```

10.23 $n = \left(\frac{(1.645)(8)}{1}\right)^2 \doteq 173.19$—take $n = 174$.

10.24 (a) $1.96\sigma/\sqrt{100} = 2.352$ points. (b) $1.96\sigma/\sqrt{10} \doteq 7.438$ points. (c) $n = \left(\frac{1.96\sigma}{3}\right)^2 \doteq 61.47$—take $n = 62$, which is under the 100-student maximum.

10.25 (a) Sources of possible error mentioned in the account are: sampling error (i.e. error due to the random nature of the sampling process), variations in the wording of questions (i.e. question bias), and variations in the order of questions asked.

(b) Only sampling (random-chance) error is covered by the announced margin of error. The other sources of error are independent of the sampling process itself; they must be controlled by the questioner, e.g., by formulating unbiased questions that do not lead the subject in a particular direction.

10.26 (a)

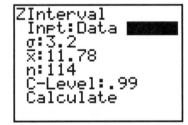 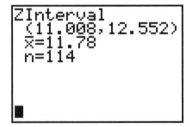

The 90% confidence interval for the mean rate of healing for newts is (22.565, 28.768).
(b)

The 99% confidence interval for the mean number of years for the hotel managers is (11.01, 12.55).

10.27 (a) $N(115, 6)$.

(b) The actual result (see facing page) lies out toward the high tail of the curve, while 118.6 is fairly close to the middle. Assuming H_0 is true, observing a value like 118.6 would not be surprising, but 125.7 is less likely, and therefore provides evidence against H_0.

(c)

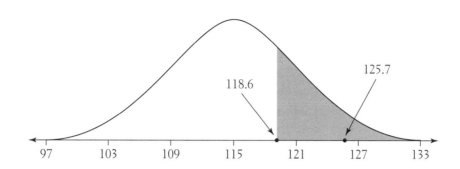

10.28 (a) N(31%, 1.518%). (b) The lower percentage lies out in the low tail of the curve, while 30.2% is fairly close to the middle. Assuming H_0 is true, observing a value like 30.2% would not be surprising, but 27.6% is unlikely, and therefore provides evidence against H_0. (c) Below.

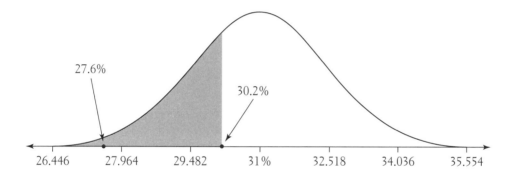

10.29 H_0: $\mu = 5$ mm; H_a: $\mu \neq 5$ mm.

10.30 H_0: $\mu = \$42{,}500$; H_a: $\mu > \$42{,}500$.

10.31 H_0: $\mu = 50$; H_a: $\mu < 50$.

10.32 H_0: $\mu = 2.6$; H_a: $\mu \neq 2.6$.

10.33 (a) The P-values are 0.2743 and 0.0373, respectively. (b) $\bar{x} = 118.6$ is significant at neither level; $\bar{x} = 125.7$ is significant at the 0.05 level, but not at the 0.01 level.

10.34 (a) The P-values are 0.2991 and 0.0125, respectively. (b) $\bar{x} = 27.6$ is significant at the 0.05 level, but not at the 0.01 level.

10.35 (a) $\bar{x} = 398$. (b) It is normal because the population distribution is normal. (c) 0.0105. (d) It is significant at $\alpha = 0.05$, but not at $\alpha = 0.01$. This is pretty convincing evidence against H_0.

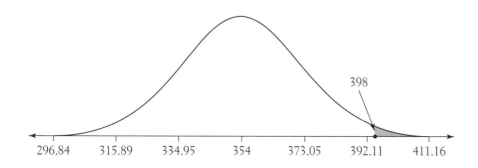

10.36 (a) Because the sample size is large (central limit theorem). (b) 0.1004. (c) Not significant at $\alpha = 0.05$. The study gives *some* evidence of increased compensation, but it is not very strong—it would happen 10% of the time just by chance.

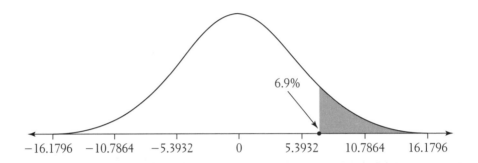

10.37 Comparing men's and women's earnings for our sample, we observe a difference so large that it would only occur in 3.8% of all samples if men and women actually earned the same amount. Based on this, we conclude that men earn more.

While there is almost certainly *some* difference between earnings of black and white students in our sample, it is relatively small—if blacks and whites actually earn the same amount, we would still observe a difference as big as what we saw almost half (47.6%) of the time.

10.38 (a) H_0: $\mu = 11.5$ vs. H_a: $\mu \neq 11.5$. (b) $z = 0.367$. (c) $P = 0.714$. This is reasonable variation when the null hypothesis is true, so we do not reject H_0.

10.39 (a) H_0: $\mu = 300$ vs. H_a: $\mu < 300$. (b) $z = -0.7893$. (c) $P = 0.2150$—this is reasonable variation when the null hypothesis is true, so we do not reject H_0.

10.40 (a) $z = -2.200$. (b) Yes, because $|z| > 1.960$. (c) No, because $|z| < 2.576$. (d) $|z|$ lies between 2.054 and 2.326. The P-value falls between 0.02 and 0.04.

10.41 (a) The command rand(100) $\rightarrow$ L$_1$ generates 100 random numbers in the interval (0,1) and stores them in list L$_1$. Here's a histogram of our simulation, and the 1-variable statistics (your results will be slightly different):

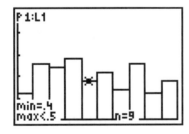 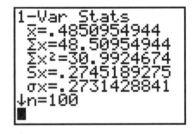

The z test statistic is $z = -.516$, and the P-value is 0.6058. We fail to reject H_0.

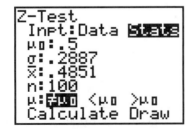 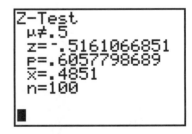

There is no evidence to suggest that the mean of the random numbers generated is different from 0.5.

10.42 (a) Yes, because $z > 1.645$. (b) Yes, because $z > 2.326$. (c) The value of z lies between 2.326 and 2.576. The P-value thus lies between 0.005 and 0.01.

10.43 (a) 99.86 to 108.41. (b) Because 105 falls in this 90% confidence interval, we cannot reject H_0: $\mu = 105$ in favor of H_a: $\mu \neq 105$.

10.44 (a) With $\bar{x} \doteq 105.84$, our 95% confidence interval for μ is $105.84 \pm (1.960)\,(15/\sqrt{31}) \doteq 105.84 \pm 5.28$, or 100.56 to 111.12 IQ points.

(b) Our hypotheses are H_0: $\mu = 100$ versus H_a: $\mu \neq 100$. Since 100 does not fall in the 95% confidence interval, we reject H_0 at the 5% level in favor of the two-sided alternative.

(c) Since the scores all came from a single school, the sample may *not* be representative of the school district as a whole.

10.45 (a) $N(0, 0.11339)$ (below). (b) $\bar{x} = 0.27$ lies out in the tail of the curve, while 0.09 is fairly close to the middle. Assuming H_0 is true, observing a value like 0.09 would not be surprising, but 0.27 is unlikely, and therefore provides evidence against H_0. (c) $P = 0.4274$ (the shaded region below). (d) $P = 0.0173$.

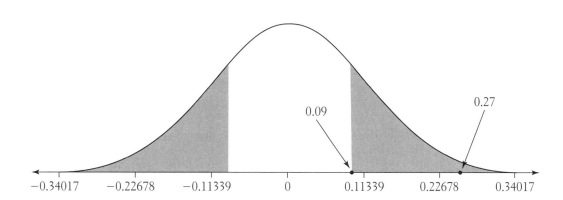

10.46 H_0: $\mu = 1250$ vs. H_a: $\mu < 1250$.

10.47 H_0: $\mu = 18$ vs. H_a: $\mu < 18$.

10.48 Hypotheses: H_0: $\mu = -0.545$ vs. H_a: $\mu > -0.545$. Test statistic: $z = 1.957$. P-value: $P = 0.0252$. We conclude that the mean freezing point really is higher, and thus the supplier *is* apparently adding water.

10.49 (a) No, because $|z| < 1.960$. (b) No, because $|z| < 1.645$.

10.50 $P = 0.1292$. Although this sample showed *some* difference in market share between pioneers with patents or trade secrets and those without, the difference was small enough that it could have arisen merely by chance. The observed difference would occur in about 13% of all samples even if there were *no* difference between the two types of pioneer companies.

10.51 Significance at the 1% level means that the P-value for the test is less than 0.01. So, it must also be less than 0.05. A result that is significant at the 5% level, by contrast, may or may not be significant at the 1% level.

10.52 The explanation is not correct; either H_0 is true (in which case the "probability" that H_0 is true equals 1) or H_0 is false (in which case this "probability" is 0). "Statistically significant at the

$\alpha = 0.05$ level" means that *if* H_0 is true, we have observed outcomes that occur less than 5% of the time.

10.53 (a) Reject H_0 if $z > 1.645$. (b) Reject H_0 if $|z| > 1.96$. (c) For tests at a fixed significance level (α), we reject H_0 when we observe values of our statistic that are so extreme (far from the mean, or other "center" of the sampling distribution) that they would rarely occur when H_0 is true. (Specifically, they occur with probability no greater than α.) For a two-sided alternative, we split the rejection region—this set of extreme values—into two pieces, while with a one-sided alternative, all the extreme values are in one piece, which is twice as large (in area) as either of the two pieces used for the two-sided test.

10.54 (a) Test H_0: $\mu = 7$ mg vs. H_a: $\mu \neq 7$ mg; since 7 is not in the interval (1.9 to 6.5 mg), we have evidence against H_0. (b) No, since 5 is in the interval.

10.55 (a) H_0: $\mu = 450$, H_a: $\mu > 450$. (b) $z = 2.46$. (c) $P = 0.007$. This represents rather strong evidence against the null hypothesis; if H_0 were true, then the observed value of $\bar{x}$ would occur in less than 1% of all samples. We therefore reject H_0. There is strong evidence against the claim that the mean math score for all California seniors is no more than 450.

10.56 (a) In Example 10.14, H_0: $\mu = 275$, H_a: $\mu < 275$, and $\sigma = 60$. Specifying a z test and entering 272 for the sample mean, the TI-83 screens that specify the information and present the results of the test are shown below. We specified "Calculate."

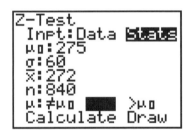 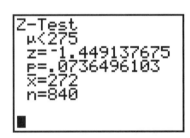

The results, $z = -1.45$ and P-value $= .0736$, agree with the results in Example 10.14.

(c) H_0: $\mu_0 = 300$, H_a: $\mu < 300$, and $\sigma = 3$. Entering the data into list L_1 and specifying a z test, here are the TI-83 screens that specify the information and present the results of the test. Again, we specified "Calculate."

The test statistic is $z = -.789$, and the P-value is 0.215. There is insufficient evidence to conclude that the mean contents of cola bottles is less than 300.

10.57 (a) $z = 1.64$; not significant at 5% level ($P = 0.0505$). (b) $z = 1.65$; significant at 5% level ($P = 0.0495$).

10.58 (a) $P = 0.3821$. (b) $P = 0.1714$. (c) $P = 0.0014$.

10.59 $n = 100$: 452.24 to 503.76. $n = 1,000$: 469.85 to 486.15. $n = 10,000$: 475.42 to 480.58.

10.60 No—the percentage was based on a voluntary response sample, and so cannot be assumed to be a fair representation of the population. Such a poll is likely to draw a higher-than-actual proportion of people with a strong opinion, especially a strong negative opinion.

10.61 (a) No—in a sample of size 500, we expect to see about 5 people who have a "P-value" of 0.01 or less. These four *might* have ESP, or they may simply be among the "lucky" ones we expect to see.

(b) The researcher should repeat the procedure on these four to see if they again perform well.

10.62 A test of significance answers question (b).

10.63 We might conclude that customers prefer design A, but perhaps not "strongly." Because the sample size is so large, this statistically significant difference may not be of any practical importance.

10.64 (a) 0.05. (b) Out of 77 tests, we can expect to see about 3 or 4 (3.85, to be precise) significant tests at the 5% level.

10.65 (a) The P-values indicate that the results observed would happen very rarely if H_0 were true; specifically, if drivers with radar detectors were the same as those without, the "before radar" speeds would be this different less than 1% of the time and "at radar" speeds would be this different less than 0.01% of the time. Since these results are so unlikely when H_0 is true, we conclude that H_0 is not true.

(b) In order for a 2-mph difference to be statistically significant (with small sample sizes), there must be little variation in the speeds; that is, those with radar detectors must all be driving at speeds very close to 70 mph, whereas those without detectors are all traveling very close to 68 mph.

10.66 (a) H_0: the patient is ill (or "the patient should see a doctor"); H_a: the patient is healthy (or "the patient should not see a doctor"). A Type I error means a false negative—clearing a patient who should be referred to a doctor. A Type II error is a false positive—sending a healthy patient to the doctor.

(b) One might wish to lower the probability of a false negative so that most ill patients are treated. On the other hand, if money is an issue, or there is concern about sending too many patients to see the doctor, lowering the probability of false positives might be desirable.

10.67 (a) Reject H_0 if $z < -2.326$. (b) 0.01 (the significance level). (c) We accept H_0 if $\bar{x} \geq 270.185$, so when $\mu = 270$, $P(\text{Type II error}) = P(\bar{x} \geq 270.185)$ $P\left(\frac{\bar{x} - 270}{60/\sqrt{840}} \geq \frac{270.185 - 270}{60/\sqrt{840}}\right) = 0.4644$.

10.68 (a) 0.50. (b) 0.1841. (c) 0.0013.

10.69 (a) Type I error: concluding that the local mean income exceeds \$45,000 when in fact it does not (and therefore, opening your restaurant in a locale that will not support it). Type II error: concluding that the local mean income does not exceed \$45,000 when in fact it does (and therefore, deciding not to open your restaurant in a locale that would support it).

(b) Type I error is more serious. If you opened your restaurant in an inappropriate area, then you would sustain a financial loss before you recognized the mistake. If you failed to open your restaurant in an appropriate area, then you would miss out on an opportunity to earn a profit there, but you would not necessarily lose money (e.g., if you chose another appropriate location in its place).

(c) H_0: $\mu = 45,000$, H_a: $\mu > 45,000$.

(d) $\alpha = 0.01$ would be most appropriate, because it would minimize your probability of committing a Type I error.

(e) In order to reject H_0 at level $\alpha = 0.01$, we must have $z \geq 2.326$, or $\bar{x} \geq 45,000 + (2.326)(5000/\sqrt{50}) = 46645$. The sample mean would have to be at least \$46,645.

(f) P(fail to reject H_0 when $\mu = 47,000$) = $P(\bar{x} < 46645$ when $\mu = 47,000)$

$$= P(Z < \tfrac{46645 - 47000}{5000/\sqrt{50}}) = P(Z < -0.5) = .3085.$$

When $\mu = 47,000$, the probability of committing a Type II error is .3085, or about 31%.

10.70 (a) Reject H_0 if $\bar{x} \geq 0.5202$. (b) 0.9666.

10.71 $z \geq 2.326$ is equivalent to $\bar{x} \geq 450 + 2.326 (100/\sqrt{500}) \doteq 460.4$, so the power is

$$P(\text{reject } H_0 \text{ when } \mu = 460) = P(\bar{x} \geq 460.4 \text{ when } \mu = 460)$$

$$= P(Z \geq \tfrac{460.4 - 460}{100/\sqrt{500}}) = P(Z \geq 0.0894) = 0.4644.$$

This is quite a bit less than the "80% power" standard; this test is not very sensitive to a 10-point increase in the mean score.

10.72 (a) Reject H_0 if $\bar{x} \leq 297.985$, so the power against $\mu = 299$ is 0.2037. (b) The power against $\mu = 295$ is 0.9926. (c) The power against $\mu = 290$ would be greater—it is further from μ_0 (300), so it is easier to distinguish from the null hypothesis.

10.73 (a) 0.5086. (b) 0.9543.

10.74 (a) Reject if $\bar{x} \geq 299.77$ or $\bar{x} \leq 250.23$. (b) Power: 0.85632. (c) $1 - 0.85632 = 0.14368$.

10.75 (a) We reject H_0 if $\bar{x} \geq 131.46$ or $\bar{x} \leq 124.54$. Power: 0.9246. (b) Power: 0.9246 (same as (a)). Over 90% of the time, this test will detect a difference of 6 (in either the positive or negative direction). (c) The power would be higher—it is easier to detect greater differences than smaller ones.

10.76 A test having low power may do a good job of not incorrectly rejecting the null hypothesis, but it is likely to accept H_0 even when some alternative is correct, simply because it is difficult to distinguish between H_0 and "nearby" alternatives.

10.77 (a) The test rejects H_0 when $|z| \geq 2.576$. The test statistic is

$$z = \frac{0.8404 - 0.86}{0.0068/\sqrt{3}} = -4.99.$$

We therefore reject H_0 (most emphatically: the P-value of the test is approximately 6 × 10^{-7}!).

(b) The test rejects H_0 when either $z \geq 2.576$ (that is, $\bar{x} \geq 0.86 + (2.576)(0.0068/\sqrt{3}) = 0.870$) or $z \leq -2.576$ (that is, $\bar{x} \leq 0.86 - (2.576)(0.0068/\sqrt{3}) = 0.850$). These are disjoint events, so the power is the sum of their probabilities, computed under the assumption that $\mu = 0.845$ is true. We have

$$P(\bar{x} \geq 0.870 \text{ when } \mu = 0.845) = P\left(Z \geq \tfrac{0.870 - 0.845}{0.0068/\sqrt{3}}\right) = P(Z \geq 6.37) \approx 0,$$

$$P(\bar{x} \leq 0.850 \text{ when } \mu = 0.845) = P\left(Z \leq \tfrac{0.850 - 0.845}{0.0068/\sqrt{3}}\right) = P(Z \leq 1.27) = 0.8980.$$

The power is thus 0.8980, or approximately 0.9.

10.78 (a) H_0: $\mu = 32$ mpg; H_a: $\mu > 32$ mpg. (b) H_0: $\mu = 24$; H_a: $\mu \neq 24$.

10.79 (a) The plot is reasonably symmetric for such a small sample. (b) 26.06 to 34.74. (c) H_0: $\mu = 25$ vs. H_a: $\mu > 25$; $z = 2.44$; P-value $= .007$. This is strong evidence against H_0.

```
2 | 034
2 |
3 | 01124
3 | 6
4 | 3
```

10.80 Confidence interval: 12.384 to 13.416. Since the sample size is relatively small, we require that the population distribution not be too nonnormal (although a sample of size 26 will overcome quite a bit of skewness). We also assume that the babies are an SRS from the population.

10.81 (a) H_0: $\mu = 32$ vs. H_a: $\mu > 32$. (b) $z = 1.8639$; P-value is 0.0312. This is strong evidence against H_0—observations this extreme would only occur in about 3 out of 100 samples if H_0 were true.

10.82 (a) Margin of error decreases. (b) The P-value decreases (the evidence against H_0 becomes stronger). (c) The power increases (the test becomes better at distinguishing between the null and alternative hypotheses).

10.83 No—"$P = 0.03$" *does* mean that the null hypothesis is unlikely, but only in the sense that the evidence (from the sample) would not occur very often if H_0 were true. P is a probability associated with the sample, not the null hypothesis; H_0 is either true or it isn't.

10.84 Yes—significance tests allow us to discriminate between random differences ("chance variation") that might occur when the null hypothesis is true, and differences that are unlikely to occur when H_0 is true.

10.85 (a) The difference observed in the study would occur in less than 1% of all samples if the two populations actually have the same proportion.

(b) The interval is constructed using a method that is correct (i.e., contains the actual proportion) 95% of the time.

(c) No—treatments were not randomly assigned, but instead were chosen by the mothers. Mothers who choose to attend a job training program may be more inclined to get themselves out of welfare.

10.86 (a) $z = \tfrac{135.2 - 115}{30/\sqrt{20}} \doteq 3.01$, which gives $P = 0.0013$. We reject H_0 and conclude that the older students do have a higher mean score.

(b) We assume the 20 students were an SRS, and that the population is (nearly) normal—near enough that the distribution of $\bar{x}$ is close to normal. The assumption that we have an SRS is more important.

10.87 $z = \frac{123.8 - 120}{10/\sqrt{40}} \doteq 2.40$, which gives $P = 0.0164$. This is strong evidence that this year's mean is different. Slight nonnormality will not be a problem since we have a reasonably large sample size (greater than 30).

10.88 (a) No treatments were imposed on the subjects (patients). This was merely an observational study.

(b) The observed association (whatever it may be) between cell phone use and the incidence of gliomas was insufficiently strong for us to conclude that it was due to an effect other than random chance. In other words, the observed association *may* have been due to a connection between cell phone use and the incidence of gliomas, but it could just as easily have been the result of chance alone.

(c) At the 5% level, one would *expect* one in every 20 tests to produce a statistically significant result by chance alone.

10.89 (a) "Exaggeration" in this case would mean that the mean breaking strength of the company's chairs was actually *less* than the claimed value of 300 pounds. Letting $\mu =$ the mean breaking strength of all chairs produced by the company, we let $H_0: \mu = 300$, $H_a: \mu < 300$.

(b) Type I error: concluding that the company's claim is exaggerated ($\mu < 300$) when in fact it is legitimate ($\mu = 300$). Type II error: concluding that the company's claim is legitimate when in fact it is invalid. Type II error would be more serious in this case, because allowing the company to continue the "false advertising" of its chairs' strength could lead to injuries, lawsuits, and other serious consequences.

(c) In order to reject H_0 at level $\alpha = 0.05$, we must have $z \leq -1.645$, or $\bar{x} \leq 300 + (-1.645)(15/\sqrt{30}) = 295.495$. All values of $\bar{x}$ at or below 295.495 pounds would cause us to reject H_0.

(d) $P(\text{Type II error}) = P(\text{fail to reject } H_0 \text{ when } \mu = 270) = P(\bar{x} > 295.495 \text{ when } \mu = 270)$

$$= P\left(Z > \frac{295.495 - 270}{15/\sqrt{30}}\right) = P(Z > 9.31) \approx 0.$$

11

Inference for Distributions

11.1 (a) $s/\sqrt{n} = 9.3/\sqrt{27} \doteq 1.7898$. (b) Since $s/\sqrt{3} = 0.01$, $s = (0.01)(\sqrt{3}) \doteq 0.0173$.

11.2 (a) 2.015. (b) 2.518.

11.3 (a) 2.145. (b) 0.688.

11.4 (a) df = 11, $t^* = 1.796$. (b) df = 29, $t^* = 2.045$. (c) df = 17, $t^* = 1.333$.

11.5 (c) The t_2 curve is a bit shorter at the peak and slightly higher in the tails (see TI-83 plot below). (d) The t_9 curve has moved toward coincidence with the standard normal curve. (e) The t_{30} curve cannot be distinguished from the standard normal curve. As the degrees of freedom increase, the t (df) curve approaches the standard normal density graph.

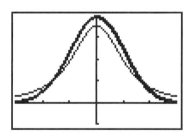

11.6 (a) 0.228.

(b), (c), and (d)

df	$P(t > 2)$	Absolute difference
2	.0917	.0689
10	.0367	.0139
30	.0273	.0045
50	.0255	.0027
100	.0241	.0013

(e) As the degrees of freedom increases, the area to the right of 2 under the t_{df} distribution gets closer to the area under the standard normal curve to the right of 2.

11.7 (a) 14. (b) 1.82 is between 1.761 ($p = 0.05$) and 2.145 ($p = 0.025$). (c) The P-value is between 0.025 and 0.05 (in fact, $P = 0.0451$). (d) $t = 1.82$ is significant at $\alpha = 0.05$ but not at $\alpha = 0.01$.

11.8 (a) 24. (b) 1.12 is between 1.059 ($p = 0.15$) and 1.318 ($p = 0.10$). (c) The P-value is between 0.30 and 0.20 (in fact, $P = 0.2738$). (d) $t = 1.12$ is not significant at either $\alpha = 0.10$ or at $\alpha = 0.05$.

11.9 (a) Since the sample size is small ($n < 15$), the distribution of the CSB vitamin C data should be close to normal. We can check this using a stemplot and a normal probability plot.

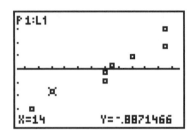

1	14
2	2236
3	11

Since there are no outliers and the normal plot is reasonably linear, the assumption of normality seems justified despite the small number of observations.

We also must assume that the eight observations represent an SRS from the population of all possible amounts of vitamin C in samples of CSB. Since the eight observations were taken from a production run, this seems like a reasonable assumption provided that the observations were taken at regular intervals.

(b) We will use the t-procedure. $\bar{x} = 22.50$, $s = 7.19$, df $= n - 1 = 7$. Using Table C with df $= 7$, we find $t^* = 2.365$. The 95% confidence interval is therefore $22.50 \pm (2.365)(7.19/\sqrt{8}) = 22.5 \pm 6.0$, or $(16.5, 28.5)$.

(c) Letting $\mu =$ the mean vitamin C content per 100 g, we wish to test $H_0: \mu = 40$ vs. $H_a: \mu \neq 40$. The t test statistic is $t = \frac{22.5 - 40}{7.2/\sqrt{8}} = -6.88$ and the corresponding P-value (from software) is 0.0002. Clearly, this result is incompatible with a process mean of $\mu = 40$. We reject H_0 and conclude that the vitamin C content for this run does not conform to specifications (specifically, it is below the specifications).

11.10 (a) The stemplot shown below has stems in 1000s, split 5 ways. The data are right-skewed with a high outlier of 2433 (and possibly 1933). The normal plot shows these two outliers, but otherwise it is not strikingly different from a line.

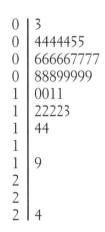

0	3
0	4444455
0	666667777
0	88899999
1	0011
1	22223
1	44
1	
1	9
2	
2	
2	4

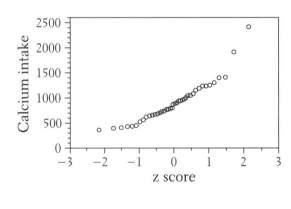

(b) $\bar{x} = 926$, $s = 427.2$, standard error $= 69.3$ (all in mg).

(c) Use of the t-procedure is justified here because the sample size is large ($n = 38 > 30$) and thus the distribution of $\bar{x}$ will be approximately normal by the central limit theorem. Using Table C with 30 degrees of freedom, we have $t^* = 2.042$. The approximate 95% confidence interval is then $926 \pm (2.042)(69.3)$, or 784.5 to 1067.5 mg; MINITAB reports 785.6 to 1066.5 mg.

(d) Without the outliers, the stemplot (see below) shows some details not previously apparent. The normal quantile plot is essentially the same as before (except that the two points that deviated greatly from the line are gone). $\bar{x} = 856.2$, $s = 306.7$, $SE_{\bar{x}} = 51.1$ (all in mg).

Using 30 degrees of freedom, we have $856.2 \pm (2.042)(51.1)$, or 751.9 to 960.5 mg; Minitab reports 752.4 to 960.0 mg.

```
 3 | 7
 4 | 01346
 5 | 47
 6 | 25789
 7 | 1478
 8 | 008
 9 | 04779
10 | 56
11 | 05
12 | 0556
13 | 2
14 | 22
```

11.11 $H_0: \mu = 1200$, $H_a: \mu < 1200$, where μ = mean daily calcium intake in mg. The t-procedure is justified for the same reason as that given in the previous exercise. The value of the t-statistic is $t = \frac{926 - 1200}{69.3} \approx -3.95$. With df = 37, we have $P = 0.00015$ (using unrounded standard error, we get $t = -3.90$ and $P = .0002$). We reject H_0 and conclude that the daily intake is significantly less than the RDA.

11.12 (a) μ is the difference between the population mean yields for Variety A plants and Variety B plants; that is, $\mu = \mu_A - \mu_B$. Another (equivalent) description is: μ is the mean difference between Variety A yields and Variety B yields.

(b) $H_0: \mu = 0$ vs. $H_a: \mu > 0$. With df = 9, we obtain $t = 1.295$, $P = 0.1137$. This is not enough evidence to reject H_0—the difference could be due to chance variation.

11.13 (a) $H_0: \mu = 0$ vs. $H_a: \mu > 0$, where μ is the mean improvement in score (posttest–pretest).

(b) Stemplot below. The stemplot of the differences, with stems split 5 ways, shows that the data are slightly left-skewed, with no outliers; the t test should be reliable.

(c) $\bar{x} = 1.450$; $SE_{\bar{x}} = 3.203/\sqrt{20} \doteq 0.716$, so $t \doteq 2.02$. With df = 19, we see that $0.025 < P < 0.05$; Minitab reports $P = 0.029$. This is significant at 5%, but not at 1%—we have some evidence that scores improve, but it is not overwhelming.

(d) Minitab gives 0.211 to 2.689; using $t^* = 1.729$ and the values of $\bar{x}$ and $SE_{\bar{x}}$ above, we obtain 1.45 ± 1.238, or 0.212 to 2.688.

```
-0 | 54
-0 | 32
-0 | 11
 0 | 11
 0 | 2223333
 0 | 4455
 0 | 7
```

11.14 (a) Neither the subjects getting the capsules nor the individuals providing them with the capsules knew which capsules contained caffeine and which were placebos.

(b) The differences between "with caffeine" and "without caffeine" heart rates for the 11 subjects are 80, 22, 17, 131, −19, 3, 23, −1, 20, −51, and −3. For these data, $\bar{x} = 20.2$, $s = 48.75$. A 95% confidence interval for the mean difference μ is $(−12.57, 52.931)$, and a test of H_0: $\mu = 0$ against H_a: $\mu \neq 0$ yields a P-value of approximately 0.2. Both statistical procedures suggest that there is no significant difference in heart rate.

11.15 (a) Methods of displaying will vary. Below is a stemplot where the digits are the stems, and all leaves are "0"—this is essentially the same as a histogram. The scores are slightly left-skewed. The normal quantile plot looks reasonably straight, except for the granularity of the data. (b) $\bar{x} = 3.618$, $s = 3.055$, $SE_{\bar{x}} = 0.524$. (c) Using df = 30, we have $t^* = 2.042$ and the interval is 2.548 to 4.688. Minitab reports 2.551 to 4.684.

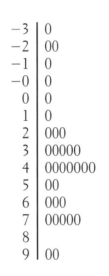

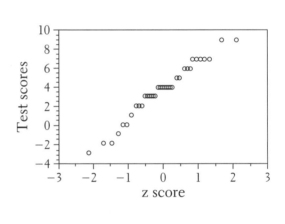

11.16 Test H_0: $\mu = 0$ vs. H_a: $\mu > 0$, where μ is the mean improvement in scores. $\pm = (\bar{x} − \mu)/SE_{\bar{x}} = 3.618/0.524 \doteq 6.90$, which has $P < 0.0005$; we conclude that scores are higher.

11.17 (a) 1.54 to 1.80. (b) We are told the distribution is symmetric; because the scores range from 1 to 5, there is a limit to how much skewness there might be. In this situation, the assumption that the 17 Mexicans are an SRS from the population is the most crucial.

11.18 (a) H_0: $\mu = 0$ vs. H_a: $\mu \neq 0$. For each subject, randomly choose which test to administer first. Alternatively, randomly assign 11 subjects to the "ARSMA first" group, and the rest to the "BI first" group. (b) $t = 4.27$; the P-value is less than 0.001, so we reject H_0. (c) 0.1292 to 0.3746.

11.19 Letting $\mu = $ the mean dimension in millimeters, we wish to test H_0: $\mu = 224$ against H_a: $\mu \neq 224$. A stem-and-leaf diagram (see next page) reveals that the distribution of the data is slightly skewed to the right, but there are no apparent outliers, so the t procedure may be used. From software, $t = 0.1254$ and $P = 0.9019$, so there is very little evidence against H_0. We have no reason to believe that the mean dimension differs from 224 mm.

```
2239 | 01
2239 | 6688899
2240 | 002
2240 | 69
2241 | 02
```

11.20 (a) The mean and standard deviation of the set of differences are $\bar{x} \doteq -5.71 \mu\text{m/hr}$ and $s \doteq 10.56 \ \mu\text{m/hr}$, so $s/\sqrt{14} \doteq 2.82 \ \mu\text{m/hr}$.

(b) We test $H_0: \mu = 0$ versus $H_a: \mu < 0$ and find $t = -5.71/2.82 \doteq -2.02$. With df $= 13$, this means that $0.025 < P < 0.05$ (software reports 0.032). This is fairly strong evidence (significant at 5% but not at 1%) that altering the electric field reduces the healing rate.

(c) Take $t^* = 1.771$; the 90% confidence interval is $-5.71 \pm (1.771)(2.82) = -10.70$ to -0.72 $\mu\text{m/hr}$. The method used to produce this interval works (gives an interval that includes the true mean) 90% of the time.

Minitab output

```
Test of mu = 0.00 vs mu < 0.00

Variable        N        Mean      StDev     SE Mean          T      P-Value
HealRate       14       -5.71      10.56        2.82      -2.02        0.032
```

11.21 (a) $H_0: \mu = 0$ vs. $H_a: \mu > 0$. $t = 43.5$; the P-value is basically 0, so we reject H_0 and conclude that the new policy would increase credit card usage.

(b) $312.14 to $351.86.

(c) The sample size is very large, and we are told that we have an SRS. This means that outliers are the only potential snag, and there are none.

(d) Make the offer to an SRS of 200 customers, and choose another SRS of 200 as a control group. Compare the mean increase for the two groups.

11.22 (a) Approximately 2.403 (from Table C), or 2.405 (using software). (b) Using $t^* = 2.403$: Reject H_0 if $t > 2.403$, which means $\bar{x} > 36.70$. (c) The power against $\mu = 100$ is 0.99998—basically 1. A sample of size 50 should be quite adequate. (d) Type I error: concluding that there has been a mean increase of $100 in the amount charged when in fact no such increase has taken place. Type II error: concluding that there has been no mean increase of $100 when in fact such an increase has taken place. Since the bank wants to be quite certain of detecting an increase, Type II error is the more serious here.

11.23 (a) The power is 0.5287. (Reject H_0 if $t > 1.833$, i.e., if $\bar{x} > 0.4811$.) (b) The power is 0.9034. (Reject H_0 if $t > 1.711$, i.e., if $\bar{x} > 0.2840$.) (c) Type I error: concluding that there is a mean difference in yield of 0.5 lb/plant when in fact there is none. Type II error: concluding that no mean difference of 0.5 lb/plant exists when in fact one does. Type II error is more serious, because the tomato experts are interested in detecting a mean difference of 0.5 when it in fact exists, and thus would prefer to use a high-power test.

11.24 (a) 9. (b) $P = 0.0255$; it lies between 0.05 and 0.025.

11.25 We want $|t| > t^*$ for $t^* =$ the upper $\alpha/2$ critical value of the t-distribution with $20 - 1 = 19$ degrees of freedom. From Table C, for $\alpha = 0.005$, this value is $t^* = 3.174$. All values t such that $|t| > 3.174$ will be statistically significant at level $\alpha = 0.005$.

11.26 We use the upper $\alpha/2$ critical value of the t-distribution with $15 - 1 = 14$ degrees of freedom. From Table C, for $\alpha = 0.02$, this value is $t^* = 2.624$.

11.27 Letting μ = the mean HAV angle in the indicated population of patients, we wish to construct a 95% confidence interval for μ. The t-procedure may be used (despite the outlier at 50) because n is large (close to 40). For these data, $\bar{x} = 25.42$, $s = 7.475$. Using 30 degrees of freedom, $t^* = 2.042$. The 95% confidence interval is $25.42 \pm (2.042) \frac{7.475}{\sqrt{38}} = 25.42 \pm 2.476$ or $(22.944, 27.896)$. MINITAB gives an interval of $(22.96, 27.88)$.

11.28 (a) Dropping the outlier at 50, we have $\bar{x} = 24.76$, $s = 6.34$. Using 30 degrees of freedom, $t^* = 2.042$ and the 95% confidence interval is $24.76 \pm (2.042) \frac{6.34}{\sqrt{37}} = 24.76 \pm 2.128$ or $(22.632, 26.888)$. MINITAB gives an interval of $(22.64, 26.87)$.

(b) The interval in part (a) is narrower than the interval in Problem 11.27. Removing the outlier decreased the standard deviation and consequently decreased the margin of error.

11.29 (a) Letting μ = the average change (Haiti − Factory) in mg/100g, the researchers wish to know if there is evidence to conclude that $\mu < 0$. Test H_0: $\mu = 0$ vs. H_a: $\mu < 0$. For the given data, $t = -4.96$ with df = 26, which corresponds to $P < 0.0005$. The mean difference is significantly less than 0, indicating that the average amount of vitamin C has decreased as a result of storage and shipment.

(b) For the change (Haiti − Factory) data, $\bar{x} = -5.33$, $s = 5.59$. For a 95% confidence interval for μ, the mean change, $t^* = 2.056$. The interval is $-5.33 \pm (2.056) \frac{5.59}{\sqrt{27}} = -5.33 \pm 2.212$ or $(-7.542, -3.118)$. MINITAB yields an interval of $(-7.54, -3.12)$.

(c) Letting μ = the mean vitamin C content of all bags shipped to Haiti, we wish to test H_0: $\mu = 40$ against H_a: $\mu \neq 40$. For the 27 Factory observations, $\bar{x} = 42.852$, $s = 4.793$. The value of the t-statistic is $t = 3.092$, which corresponds to a P-value of between 0.002 and 0.005 (two-sided case). There is strong evidence indicating that μ in fact differs from the target mean of 40 mg/100g.

Minitab output

```
Test of mu = 40.000 vs mu not = 40.000

Variable      N      Mean     StDev    SE Mean      T          P
VITC         27    42.852     4.793     0.923     3.09     0.0047
```

11.30 (a) 109.97 to 119.87. (b) We assume that the 27 members of the placebo group can be viewed as an SRS of the population, and that the distribution of seated systolic BP in this population is normal, or at least not too nonnormal. Since the sample size is somewhat large, the procedure should be valid as long as the data show no outliers and no strong skewness.

11.31 (a) Randomly assign 12 (or 13) into a group that will use the right-hand knob first; the rest should use the left-hand knob first. Alternatively, for each student, randomly select which knob he or she should use first.

(b) μ is the mean difference between right-handed times and left-handed times; the null hypothesis is H_0: $\mu = 0$ (no difference). How the alternative is written depends on exactly how μ is defined. Let μ_R be the mean right-hand thread time for all right-handed people (or students), and μ_L be the mean left-hand thread time. As described above, we would most naturally write $\mu = \mu_R - \mu_L$; in this case, H_a: $\mu < 0$. Alternatively, we might define $\mu = \mu_L$

$- \mu_R$, so that H_a: $\mu > 0$. Either way, the null hypothesis says $\mu_R = \mu_L$ and the alternative is $\mu_R < \mu_L$.

(c) A plot of the differences shows no outliers or strong skewness. $\bar{x} = -13.32$ (or $+13.32$), $SE(\bar{x}) = 4.5872$, $t = \pm 2.9037$, and $P = 0.0039$. We reject H_0 in favor of H_a.

11.32 5.47 to 21.17 seconds. For our sample $\bar{x}_R \div \bar{x}_L = 88.7\%$; this suggests that right-handed students working on an assembly line with right-handed threads would complete their task in about 90% of the time that it would take them to complete the same task with left-handed threads.

11.33 (a) See below. (b) H_0: $\mu = 105$ vs. H_a: $\mu \neq 105$, $t = -0.3195$, $P = 0.7554$. We do not reject the null hypothesis—the mean detector reading could be 105.

$$
\begin{array}{r|l}
9 & 2 \\
9 & 578 \\
10 & 024 \\
10 & 55 \\
11 & 1 \\
11 & 9 \\
12 & 2 \\
\end{array}
$$

11.34 We know the data for *all* presidents; we know about the whole population, not just a sample. (We might want to try to make statements about future presidents, but doing so from this data would be highly questionable; they can hardly be considered an SRS from the population.)

11.35 (a) 2.080. (b) Reject H_0 if $|t| \geq 2.080$, i.e., if $|\bar{x}| \geq 0.133$. (c) $P(|\bar{x}| \geq 0.133) = P(\bar{x} \leq -0.133$ or $\bar{x} \geq 0.133) = P(Z \leq -5.207$ or $Z \geq -1.047) = 0.852$.

11.36 (a) Yes, provided that the sample is representative of the population at large. (For example, the papers should not all come from the same school or class.) Since the population of 42,000 is much larger than the sample size of 25, the population size itself does not play much of a role in the variability of the sampling distribution of $\bar{x}$.

(b) Probably not. The 25 observations have a strongly right-skewed distribution, indicating that the scores on this question tend to be lower than the "average" value of 2.

(c) While the data are skewed, the absence of outliers and the moderately large sample size ($n = 25$) will permit us to use the t-procedure as an approximation. For these data, $\bar{x} = 1.04$ and $s = 1.14$. The critical value t^* (with 24 degrees of freedom) $= 2.064$, and the 95% confidence interval is $1.04 \pm (2.064)\frac{1.14}{\sqrt{25}} = 1.04 \pm 0.471$ or $(0.569, 1.511)$. MINITAB yields an interval of $(0.571, 1.509)$.

Minitab output

Variable	N	Mean	StDev	SE Mean	95.0 % CI
APSCORE	25	1.040	1.136	0.227	(0.571, 1.509)

11.37 (a) (3)—two samples. (b) (2)—matched pairs.

11.38 (a) (1)—single sample. (b) (3)—two samples.

11.39 Both sample sizes are quite large, so we do not need to worry about the normality of the corresponding populations. Letting $\mu_1 =$ the mean Chapin Test score for males and $\mu_2 =$ the mean Chapin Test score for females, we wish to test H_0: $\mu_1 = \mu_2$ against H_a: $\mu_1 \neq \mu_2$. The two-sample t statistic for this test is $t = \dfrac{25.34 - 24.94}{\sqrt{\frac{(5.05)^2}{133} + \frac{(5.44)^2}{162}}} = 0.654$. df $= 132$, so use the $t(100)$ distribution in Table C: we find that $P > 0.50$. The data provide no evidence that males and females differ in social insight.

11.40 (a) If the loggers had known that a study would be done, they might have (consciously or subconsciously) cut down fewer trees than they typically would, in order to reduce the impact of logging.

(b) We test H_0: $\mu_1 = \mu_2$ versus H_a: $\mu_1 > \mu_2$. The means and standard deviations are given in the Minitab output below; we compute SE $= \sqrt{3.53^2/12 + 4.50^2/9} \doteq 1.813$ and $t = \frac{17.50 - 13.67}{1.813} \doteq 2.11$. With df $= 8$, we find $0.025 < P < 0.05$; this is significant at 5% but not at 1%.

(c) Use $t^* = 1.860$: $(17.50 - 13.67) \pm (1.860)(1.813) = 0.46$ to 7.20.

Minitab output

```
Twosample T for Species
Code    N     Mean    StDev    SE Mean
1      12    17.50     3.53       1.0
2       9    13.67     4.50       1.5
```

11.41 (a) The study was a randomized comparitive experiment. The large sample sizes will help ensure the accuracy of the t-procedures. Test H_0: $\mu_1 = \mu_2$ against H_a: $\mu_1 < \mu_2$ where μ_1 and μ_2 are the mean lengths of stay for normothermic (blanket) and hypothermic (non-blanket) patients. $t = \dfrac{12.1 - 14.7}{\sqrt{\frac{(4.4)^2}{104} + \frac{(6.5)^2}{96}}} = -3.285$. Using df $= 80$ in Table C, we find that $0.0005 < P < 0.01$. Heating blanketsdo seem to reduce the length of a patient's hospital stay.

(b) Using df $= 80$, the critical value $t^* = 1.990$ and the confidence interval for $\mu_1 - \mu_2$ is $(12.1 - 14.7) \pm (1.990) \sqrt{\frac{(4.4)^2}{104} + \frac{(6.5)^2}{96}} = -2.6 \pm 1.575$ or $(-4.175, -1.025)$. Since the interval contains only negative values, it reinforces the result of the test in (a) that heating blankets reduce the lengths of stays. Specifically, we are 95% confident that the mean reduction is between (roughly) 1 and 4 days.

11.42 (a) Because the sample sizes are so large (and the sample sizes are almost the same), deviation from the assumptions have little effect.

(b) Using $t^* = 1.660$ from a $t(100)$ distribution, the interval is \$412.68 to \$635.58. Using $t^* = 1.6473$ from a $t(620)$ distribution (obtained with software), the interval is \$413.54 to \$634.72.

(c) The sample is not *really* random, but there is no reason to expect that the method used should introduce any bias into the sample.

(d) Students without employment were excluded, so the survey results can only (possibly) extend to *employed* undergraduates. Knowing the number of unreturned questionnaires would also be useful.

11.43 H_0: $\mu_1 = \mu_2$ vs. H_a: $\mu_1 > \mu_2$, where μ_1 and μ_2 are the mean number of beetles on untreated (control) plots and malathion-treated plots, respectively. $t = 5.8090$, which yields $P < 0.0001$ for a $t(12)$ distribution—this is significant at the 1% level.

11.44 Consider the left and right endpoint of the confidence interval. If these endpoints remain fixed, then as the degrees of freedom increase, the area in the tails (outside the confidence interval and under the t_{df} curve) decreases. See Exercise 11.5. To make up this difference in area, the endpoints of the intervals have to move toward the center of the distribution, giving up some area to the tails. Thus the confidence interval becomes narrower.

11.45 Here are the key calculator screens.

11.46 The test statistic is $t = 2.99$. With df $= 5.9376$, the P-value is .0246.

11.47 (a) H_0: $\mu_{\text{skilled}} = \mu_{\text{novice}}$ vs. H_a: $\mu_s > \mu_n$. (b) The t statistic we want is the "Unequal" value: $t = 3.1583$; its P-value is 0.0052. This is strong evidence against H_0. (c) Using $t^* = 1.895$ from a $t(7)$ distribution: 0.4691 to 1.876. Using $t^* = 1.8162$ from a $t(9.8)$ distribution (from software): 0.4982 to 1.8474.

11.48 H_0: $\mu_{\text{skilled}} = \mu_{\text{novice}}$ vs. H_a: $\mu_s \ne \mu_n$ (use a two-sided alternative since we have no preconceived idea of the direction of the difference). The t statistic we want is $t = 0.5143$; its P-value is 0.6165. There is no significant difference in weight between skilled and novice rowers.

11.49 (a) With the small sample sizes, and with no way to check the normality of the data, the t-procedures will be only approximately accurate. Test H_0: $\mu_1 = \mu_2$ against H_a: $\mu_1 \ne \mu_2$, where μ_1 and μ_2 are the mean number of trials to completion for the nicotine group and the saline (control) group. $t = \dfrac{111.9 - 75.6}{\sqrt{\frac{(50.28)^2}{10} + \frac{(27.512)^2}{10}}} = 2.003$. Using df $= 9$ in Table C, $0.05 < P < 0.1$.

Using the TI-83, the more precise df $= 13.946$ and $P = 0.065$. There does not seem to be very strong evidence of a difference between the two means.

(b) Using df $= 9$, the conservative degrees of freedom, the critical value $t^* = 2.262$ and the 95% confidence interval for $\mu_1 - \mu_2$ is $(111.9 - 75.6) \pm (2.262) \sqrt{\frac{(50.28)^2}{10} + \frac{(27.512)^2}{10}} = 36.3 \pm 41$ or $(-4.7, 77.3)$. Using df $= 13.946$, the TI-83 gives the 95% confidence interval as $(-2.587, 75.187)$. The more precise df results in a slightly narrower interval.

11.50 (a) Two-sample t test. (b) Matched pairs t test. (c) Matched pairs t test. (d) Two-sample t test. (e) Matched pairs t test.

11.51 (a) Completed table below. The only values not given directly are the standard deviations, which are found by computing $s = \text{SEM}\sqrt{10}$. (b) Use df $= 9$ (the "smaller of 9 and 9").

Treatment	n	$\bar{x}$	s
IDX	10	116	17.71
Untreated	10	88.5	6.01

11.52 (a) The means and SEMs were given in Exercise 11.51 as 88.5 and 1.9 days (control) and 116 and 5.6 days (IDX), so $SE = \sqrt{1.9^2 + 5.6^2} \doteq 5.91$ and $t = (88.5 - 116)/SE \doteq -4.65$. With either df = 9 or df = 11.04, we have a significant result ($P < 0.001$ or $P < 0.0005$, respectively), so there is strong evidence that IDX prolongs life.

(b) The interval is $(116 - 88.5) \pm t^*(5.91)$, which equals either 14.13 to 40.87 days (df = 9, $t^* = 2.262$) or 14.49 to 40.51 days (df = 11, $t^* = 2.201$).

11.53 (a) Stemplots show little skewness, but one moderate outlier (85) for the control group on the right. Nonetheless, the t procedures should be fairly reliable since the total sample size is 44.

(b) $H_0: \mu_t = \mu_c$ vs. $H_a: \mu_t < \mu_c$; $t = 2.311$. Using $t(20)$ and $t(37.9)$ distributions, P equals 0.0158 and 0.0132, respectively; reject H_0.

(c) Randomization was not really possible, because existing classes were used—the researcher could not shuffle the students.

```
         | 1 | 079
       4 | 2 | 068
       3 | 3 | 377
 9964333 | 4 | 1222368
98776432 | 5 | 3455
     721 | 6 | 02
       1 | 7 |
         | 8 | 5
```

11.54 (a) The observations are "before-and-after" weights, so the pairs of observations will be highly correlated—it is the change in weight that we are interested in.

(b) We expect some variation in the weight change, and there may have been some loss due to chance, but the amount lost was so great that it is unlikely to occur merely by chance. In short, this weight-loss program seems to work.

(c) Table C shows that the P-value must be smaller than 0.0005; in fact, it is less than 0.00002.

11.55 (a) A stemplot of the differences (below) looks reasonably normal with no outliers, so the t procedures should be safe.

(b) For the differences, $SE = s/\sqrt{n} = 1071/\sqrt{20} \approx 239$. With $df = 19$, $t^* = 1.729$ and the interval is $-37 \pm (1.729)(239)$, or -451 to 376.5.

```
-2 | 0
-1 | 6
-1 | 1
-0 | 775
-0 | 433210
 0 | 123
 0 | 5
 1 | 13
 1 | 5
 2 | 3
```

11.56 (a) A matched pairs t test should be used because the observations from Try 1 and Try 2 are dependent; each coached student generates a pair of observations.

(b) The t-procedures may be used because the sample size is very large. For μ = mean gain (Try 2 score − Try 1 score), we will test $H_0: \mu = 0$ vs. $H_a: \mu > 0$. The test statistic is

$t = \frac{29}{59/\sqrt{427}} = 10.157$. Using df $= 100$ in Table C, we obtain $P < 0.0005$. Using the TI-83, we get $P = 3.77 \times 10^{-22} \approx 0$. Coaching definitely seems to have improved the students' scores.

(c) $(21.612, 36.388)$.

11.57 (a) Since the sample sizes are quite large, we should be safe applying the t-procedures. Test H_0: $\mu_1 = \mu_2$ against H_a: $\mu_1 > \mu_2$, where μ_1 and μ_2 are the mean gains for coached and uncoached students respectively, in a two-sample t test. $t = \dfrac{29 - 21}{\sqrt{\frac{(59)^2}{427} + \frac{(52)^2}{2733}}} = 2.646$.

Using df $= 100$ in Table C, we find that $0.0025 < P < 0.005$. The TI-83 yields $P = 0.0042$. There is evidence that coached students gained more on the average than uncoached students.

(b) Using the conservative df $= 100$, the critical value $t^* = 2.626$ and the confidence interval is $(29 - 21) \pm (2.626)\sqrt{\frac{(59)^2}{427} + \frac{(52)^2}{2733}} = 8 \pm 7.94$ or $(0.06, 15.94)$. Using the TI-83, the more precise confidence interval is $(0.184, 15.816)$. The intervals suggest that on the average, coached students gained somewhere between 0 and 15 points more than uncoached students.

(c) The average gain vis-a-vis uncoached students is rather small, so it may well be that coaching courses are *not* worth the time and expense.

11.58 Since the same students are taking the test over again, the effect of coaching will be confounded with the "experience factor," i.e., the tendency of a student to improve his/her score on Try 2 because of familiarity with the format and types of questions asked on the test. It is impossible to separate the two effects, so a "cause-and-effect" relationship cannot be inferred.

11.59 E.g., The difference between average female (55.5) and male (57.9) self-concept scores was so small that it can be attributed to chance variation in the samples ($t = -0.83$, df $= 62.8$, $P = 0.4110$). In other words, based on this sample, we have no evidence that mean self-concept scores differ by gender.

11.60 (a) $t^* = 2.364$, the value for a $t(100)$ distribution (since values for a $t(99)$ distribution are not given). (b) Reject H_0 when $\bar{x}_1 - \bar{x}_2 \geq 2.6746$. (c) Power: $P(Z \geq -2.0554) = 0.9801$. (d) Type I error: detecting a lowering of blood pressure when none exists. Type II error: failing to detect a lowering of blood pressure when such an effect does exist. Since the clinical study is primarily interested in detecting a "beneficial" effect if it in fact exists, Type II error should be considered the more serious.

11.61 (a) H_0: $\mu_A = \mu_B$ vs. H_a: $\mu_A \neq \mu_B$; $t = (\bar{x}_A - \bar{x}_B)/\sqrt{\dfrac{s_A^2}{n_A} + \dfrac{s_B^2}{n_B}}$.

(b) For a $t(349)$ distribution, $t^* = 1.967$; using a $t(100)$ distribution, take $t^* = 1.984$.

(c) We reject H_0 when $|\bar{x}_A - \bar{x}_B| \geq 59.48$ (using $t^* = 1.967$). To find the power against $|\mu_A - \mu_B| = 100$, we choose *either* $\mu_A - \mu_B = 100$ or $\mu_A - \mu_B = -100$ (the probability is the same either way). Taking the former, we compute: $P[(\bar{x}_A - \bar{x}_B) \leq -59.48$ or $(\bar{x}_A - \bar{x}_B) \geq 59.48] = P(Z \leq -5.274$ or $Z \geq -1.340) = 0.9099$. Repeating these computations with $t^* = 1.984$ gives power 0.9071.

(d) Type I error: detecting a difference of $100 in mean amount charged when no such difference exists. Type II error: failing to detect a difference of $100 in mean amount charged when the difference does exist. Type II error should be of more concern to the bank because if there is a significant difference in the mean amount charged, the bank needs to know about it and adjust the handling of credit-card accounts appropriately.

11.62 (a) This is a two-sample t test—the two groups of women are (presumably) independent. (b) Use a $t(44)$ distribution. (c) The sample sizes are large enough that nonnormality has little effect on the reliability of the procedure.

11.63 (a) (See stemplot below.) The distribution looks reasonably symmetric; other than the low (9.4 ft) and high (22.8 ft) outliers, it appears to be nearly normal. The mean is $\bar{x} \doteq 15.59$ ft and the standard deviation is $s = 2.550$ ft.

(b) Using df = 40 from Table C, $t^* = 2.021$, so the interval is $\bar{x} \pm t^* s / \sqrt{44} \doteq 15.59 \pm 0.78 = 14.81$ to 16.37 ft. Using software, we can find $t^* = 2.0167$ for df = 43, which also (after rounding) gives 15.59 ± 0.78 ft. Since 20 ft does not fall in (or even near) this interval, we reject this claim.

(c) We need to know what population we are examining: Were these all full-grown sharks? Were they all male? (I.e., is μ the mean adult male shark length? Or something else?)

```
 9 | 4
10 |
11 |
12 | 12346
13 | 22225668
14 | 3679
15 | 237788
16 | 122446788
17 | 688
18 | 23677
19 | 17
20 |
21 |
22 | 8
```

11.64 (a) Matched pairs t test. (b) Two-sample t test. (c) Two-sample t test. (d) Matched pairs t test. (e) Matched pairs t test.

11.65 (a) Side-by-side boxplots of the data (see below) reveal rough symmetry for the air-filled football and strong left-skewness for the helium-filled football.

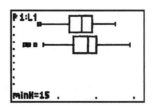

A two-sample t test of H_0: $\mu_A = \mu_H$ vs. H_a: $\mu_A < \mu_H$ yields $t = -0.37$, $P = 0.356$. We fail to reject H_0; there is no significant difference in the lengths of the kicks for the two balls.

(b) Without the outliers (i.e., the values 11, 12, 14, and 14 in the helium data set), $t = -1.931$ and $P = 0.0287$. Now we might conclude that there is a significant difference in the mean distance traveled by air-filled and helium-filled footballs.

(c) A time plot for each of the two balls shows an increasing trend, but there are also occasional short kicks that deviate from the trend. You could compare the lengths of the first 20 kicks and the remaining 19 kicks with each ball. For the air-filled ball, side-by-side boxplots of the first 20 kicks and the remaining 19 kicks (see below) suggest that the kicker may have improved. A two-sample t test for the two sets of "air-filled data" yields $t = -1.736$ and $P = 0.046$.

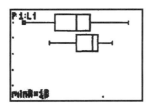

11.66 (a) The "unscented first" and "scented first" groups were separate, not matched pairs.

(b) The stemplot (below) and the means ($\bar{x}_1 = 54.38$ sec and $\bar{x}_2 = 45.21$ sec) suggest that unscented-first times are longer; that is, those subjects were slower. The unscented-first data show some hint of nonnormality, but the t procedures should be safe.

(c) We test H_0: $\mu_1 = \mu_2$ versus H_a: $\mu_1 > \mu_2$. The standard deviations are $s_1 \doteq 17.49$ and $s_2 \doteq 8.348$ sec, so $SE \doteq 5.897$ and $t \doteq 1.55$. With either df $= 9$ or df $= 14.6$, this is not significant ($0.05 < P < 0.10$), so we don't have strong enough evidence to conclude that there was a learning effect.

Unscented 1st		Scented 1st
710	3	267
3	4	3378
43	5	138
8540	6	
	7	
7	8	

11.67 (a) Using a $t(1361)$ distribution, you get \$1016.56 to \$1069.44, using a $t(2669.1)$ distribution, you get almost the same interval: \$1016.58 to \$1069.42. (b) Skewness will have little effect because the sample sizes are very large.

11.68 (a) First compute each subject's improvement ("after" minus "before"). We test H_0: $\mu_1 = \mu_2$ versus H_a: $\mu_1 > \mu_2$. The means and standard deviations are given in the Minitab output below; we compute $SE = \sqrt{3.17^2/10 + 3.69^2/8} \doteq 1.645$ and $t = \frac{11.40 - 8.25}{1.645} \doteq 1.91$. With df $= 7$, we find $0.025 < P < 0.05$; this is significant at 5% (and also at 10%). (Note that Minitab uses the more accurate df $= 13$, rather than the conservative approach.) (b) Use $t^* = 1.895$: $(11.40 - 8.25) \pm (1.895)(1.645) = 0.03$ to 6.27. (Minitab's result is based on df $= 13$, so is slightly narrower than the conservative interval.)

Minitab output

```
Two sample T for Diff
Code     N      Mean    StDev    SE Mean
1       10     11.40     3.17      1.0
2        8      8.25     3.69      1.3

90% C.I. for mu 1 − mu 2: (0.2, 6.1)
T-Test mu 1 = mu 2 (vs >): T = 1.91 P=0.039 DF = 13
```

11.69 We can compare heart rates between the groups by conducting two-sample t tests of the form $H_0: \mu_1 = \mu_2$ vs. $H_a: \mu_1 \neq \mu_2$. There are three separate tests to conduct: control group (C) vs. friend group (F), control group (C) vs. pet group (P), and friend group (F) vs. pet group (P). Below are the MINITAB outputs for these three cases. (In each case, the precise value of df is being used.)

```
Two sample T for CONTROL vs FRIEND

               N      Mean    StDev    SE Mean
CONTROL       15     106.5     10.1      2.6
FRIEND        15     117.7     13.0      3.4

95% CI for mu CONTROL - mu FRIEND: (−19.9, −2.4)
T-Test mu CONTROL = mu FRIEND (vs not =): T = −2.62 P = 0.015
DF = 26
```

There is evidence of a significant difference between the mean heart rates of the control group and the friend group. Specifically, since the 95% confidence interval contains only negative values, the mean for the control group appears to be significantly *lower* than the mean for the friend group (difference in means = 11.2).

```
Two sample T for CONTROL vs PET

               N      Mean    StDev    SE Mean
CONTROL       15     106.5     10.1      2.6
PET           15      78.5     15.4      4.0

95% CI for mu CONTROL - mu PET: (18.2, 37.8)
T-Test mu CONTROL = mu PET (vs not =): T = 5.90 P = 0.0000 DF = 24
```

There is very strong evidence of a significant difference between the control group and the pet group. In this case, since the 95% confidence interval contains only positive values, the mean for the control group is significantly *higher* than the mean for the pet group (difference in means = 28.0).

```
Two sample T for FRIEND vs PET

               N      Mean    StDev    SE Mean
FRIEND        15     117.7     13.0      3.4
PET           15      78.5     15.4      4.0

95% CI for mu FRIEND - mu PET: (28.5, 49.8)
T-Test mu FRIEND = mu PET (vs not =): T = 7.52 P = 0.0000 DF = 27
```

There is very strong evidence of a significant difference between the friend group and the pet group. Since the 95% confidence interval contains only positive values, the mean for the friend group is significantly *higher* than the mean for the pet group (difference in means = 39.2).

Combining the results of the tests, it appears that the friend group has the highest heart rate, the control group has the second highest, and the pet group has the lowest.

11.70 No—you have information about all Indiana counties (not just a sample).

11.71 The stemplot looks fairly symmetric; 4.88 is perhaps a moderate low outlier, but is not too far from the other observations. Our estimate is the mean, $\bar{x} = 5.4479$. The standard error of the mean is 0.0410; the margin of error depends on the confidence level chosen. Here are three possibilities:

Confidence level	Confidence interval	Margin of error
90%	(5.3781,5.5177)	0.0698
95%	(5.3639,5.5320)	0.0840
99%	(5.3345,5.5613)	0.1134

```
48 | 8
49 |
50 | 7
51 | 0
52 | 6799
53 | 04469
54 | 2467
55 | 03578
56 | 12358
57 | 59
58 | 5
```

11.72 (a) $H_0: \mu_1 = \mu_2$ vs. $H_a: \mu_1 > \mu_2$; $t = 1.1738$, so $P = 0.1265$ (using $t(22)$) or 0.123453 (using $t(43.3)$). Not enough evidence to reject H_0. (b) -14.57 to 52.57 (using df = 22), or -13.64 to 51.64 (using df = +3.3). (c) 165.53 to 220.47. (d) We are assuming that we have two SRSs from each population, and that underlying distributions are normal. It is unlikely that we have random samples from either population, especially among pets.

11.73 (a) Stemplots (see below) show that both distributions are skewed right. There is one high outlier in the "Active" group. The summary statistics are:

	n	$\bar{x}$	s
Active	24	24.4167	6.31022
Passive	24	17.8750	4.02506

To test $H_0: \mu_A = \mu_P$ versus $H_a: \mu_A > \mu_P$, we find that $SE \doteq 1.5278$ and $t \doteq 4.28$. With either df = 23 or df = 39.1, $P < 0.0005$, so there is strong evidence that active learning results in more correct identifications. (b) With $t^* = 1.714$ from a $t(23)$ distribution, the 90% confidence interval is $24.4167 \pm (1.714)(6.31022/\sqrt{24}) = 22.2$ to 26.6 Blissymbols. (Note: This is a one-sample question.)

Active		Passive
	1	223
5	1	45555
76	1	66777
	1	889
111100	2	00111
332	2	
4444	2	5
7	2	66
9888	2	
1	3	
	3	
5	3	
	3	
	3	
	4	
	4	
4	4	

12

Inference for Proportions

12.1 (a) Population: the 175 residents of Tonya's dorm; p is the proportion who like the food. (b) $\hat{p} = 0.28$.

12.2 (a) The population is the 2400 students at Glen's college, and p is the proportion who believe tuition is too high. (b) $\hat{p} = 0.76$.

12.3 (a) The population is the 15,000 alumni, and p is the proportion who support the president's decision. (b) $\hat{p} = 0.38$.

12.4 (a) No—the population is not large enough relative to the sample. (b) Yes—we have an SRS, the population is 48 times as large as the sample, and the success count (38) and failure count (12) are both greater than 10. (c) No—there were only 5 or 6 "successes" in the sample.

12.5 (a) No—np_0 and $n(1 - p_0)$ are less than 10 (they both equal 5). (b) No—the expected number of failures is less than 10 ($n(1 - p_0) = 2$). (c) Yes—we have an SRS, the population is more than 10 times as large as the sample, and $np_0 = n(1 - p_0) = 10$.

12.6 (a) $SE_{\hat{p}} = \sqrt{(0.54)(0.46)/1019} \doteq 0.01561$, so the 95% confidence interval is $0.54 \pm (1.96)(0.01561) = 0.51$ to 0.57. The margin of error is about 3%, as stated.

(b) We weren't given sample sizes for each gender. (However, students who know enough algebra can get a good estimate of those numbers by solving the system $x + y = 1019$ and $0.65x + 0.43y = 550$: approximately 508 men and 511 women.)

(c) The margin of error for women alone would be greater than 0.03 since the sample size is smaller.

12.7 (a) The methods can be used here, since we assume we have an SRS from a large population, and all relevant counts are more than 10. For TVs in rooms: $\hat{p}_1 \doteq 0.66$ and $SE_{\hat{p}} = \sqrt{(0.66)(0.34)/1048} \doteq 0.01463$, so the 95% confidence interval is $0.66 \pm (1.96)(0.01463) \doteq 0.631$ to 0.689. For preferring Fox: $\hat{p}_2 \doteq 0.18$ and $SE_{\hat{p}} = \sqrt{(0.18)(0.82)/1048} \doteq 0.01187$, so the 95% confidence interval is $0.18 \pm (1.96)(0.01187) \doteq 0.157$ to 0.203.

(b) In both cases, the margin of error for a 95% confidence interval ("19 cases out of 20") was (no more than) 3%.

(c) We test H_0: $p = 0.5$ versus H_a: $p > 0.5$. The test statistic is $z = (0.66 - 0.50)/\sqrt{\frac{(0.5)(0.5)}{1048}} \doteq 10.36$, which gives very strong evidence against H_0 ($P < 0.0002$); we conclude that more than half of teenagers have TVs in their rooms. (Additionally, the interval from (a) does not include 0.50 or less.) With the TI-83, $z = 10.379$ and $P = 1.577 \times 10^{-25}$.

12.8 (a) $\hat{p} = .66$, and since $n\hat{p} = 132$ and $n(1 - \hat{p}) = 68$ are both greater than 10, the confidence interval based on z can be used. The 95% confidence interval for p is $.66 \pm (1.96)\sqrt{((.66)(.34)/200)} = .66 \pm 0.06565$, or 0.59435 to 0.72565.

(b) Yes; the 95% confidence interval contains *only* values that are less than 0.73, so it is likely that for this particular population, p differs from 0.73 (specifically, is less than 0.73).

12.9 (a) $\hat{p} = \frac{15}{84} \doteq 0.1786$, and $SE_{\hat{p}} = \sqrt{\hat{p}(1-\hat{p})/84} \doteq 0.0418$. (b) Checking conditions, $n\hat{p} = 15$ and $n(1-\hat{p}) = 69$ are both at least 10. Provided that there are at least $(84)(p) = 840$ applicants in the population of interest, we are safe constructing the confidence inerval. $\hat{p} \pm 1.645\ SE_{\hat{p}} = 0.1098$ to 0.2473.

12.10 $n = \left(\frac{1.645}{0.04}\right)^2(0.7)(0.3) \doteq 355.2$—use $n = 356$. With $\hat{p} = 0.5$, $SE_{\hat{p}} \doteq 0.0265$, so the true margin of error is $(1.645)(0.0265) = 0.0436$.

12.11 (a) 1051.7—round up to 1052. (b) 1067.1—round up to 1068; 16 additional people.

12.12 450.2—round up to 451.

12.13 (a) We do not know that the examined records came from an SRS, so we must be cautious in drawing emphatic conclusions. Both $n\hat{p}$, $n(1-\hat{p})$ are at least 10. $\hat{p} = \frac{542}{1711} \doteq 0.3168$; $SE_{\hat{p}} = \sqrt{\hat{p}(1-\hat{p})/1711} \doteq 0.01125$; the interval is $\hat{p} \pm 1.960\ SE_{\hat{p}} = 0.2947$ to 0.3388. (b) No: We do not know; for example, what percentage of cyclists who were *not* involved in fatal accidents had alcohol in their systems.

12.14 $\hat{p} = 0.05227$, $z = -3.337$, and $P < 0.0005$—very strong evidence against H_0, and in favor of $H_a: p < \frac{1}{10}$.

12.15 (a) Checking conditions: we are treating the Gallup sample as an SRS; the population of adults is much larger than 10(1785); $n\hat{p}$ and $n(1-\hat{p})$ are both at least 10. The interval is 39.0% to 45.0%. (b) Since 50% falls above the 99% confidence interval, this is strong evidence against H_0: $p = 0.5$ in favor of $H_a: p < 0.5$. (In fact, $z = -6.75$ and P is tiny.) (c) 16589.4—round up to 16590. The use of $p^* = 0.5$ is reasonable because our confidence interval shows that the actual p is in the range 0.3 to 0.7.

12.16 (a) The distribution is approximately normal with mean $\mu = p = 0.14$ and standard deviation $\sigma = \sqrt{(0.14)(0.86)/9224} \doteq 0.003613$.

   (b) For testing $H_0: p = 0.14$ versus $H_a: p > 0.14$, we have $\hat{p} \doteq 0.27$, and the test statistic is $z = (0.27 - 0.14)/0.003613 \doteq 36$. We have very strong (in fact, overwhelming) evidence that Harleys are more likely to be stolen.

12.17 (a) $H_0: p = 0.5$ vs. $H_a: p > 0.5$, $z = 1.697$, $P = 0.0448$—reject H_0 at the 5% level. (b) 0.5071 to 0.7329. (c) The coffee should be presented in random order—some should get the instant coffee first, and others the fresh-brewed first.

12.18 (a) $n = \left(\frac{2.576}{0.015}\right)^2(0.2)(0.8) \doteq 4718.8$—use $n = 4719$. (b) $2.576\sqrt{\frac{(0.1)(0.9)}{4719}} \doteq 0.01125$.

12.19 (a) Let $p =$ Shaq's free-throw percentage during the season following his off-season training. We wish to test $H_0: p = .533$ vs. $H_a: p > .533$. $np_0 = (39)(.533) = 27.087$ and $n(1 - p_0) = (39)(.467) = 18.213$ are both greater than 10, so the one-sample z test may be used. The test statistic is $z = (.667 - .533)/\sqrt{((.533)(.467)/39)} = 1.673$, and the P-value $= 0.04715$. There is some evidence that Shaq has, in fact, improved his free-throw percentage.

   (b) Type I error: concluding that Shaq has improved his free-throwing when in fact he has not. Type II error: concluding that Shaq has not improved his free-throwing when in fact he has.

   (c) We seek the power of the test when $p = 0.6$. With $\alpha = 0.05$, we reject H_0 in favor of H_a when $z > 1.645$, that is, when $\hat{p} > 0.533 + (1.645)(\sqrt{((.533)(.467)/39)}) = 0.6644$. Then, $P(\text{reject } H_0 \text{ when } p = 0.6) = P(\hat{p} > 0.6644 \text{ when } p = 0.6) = P(Z > 0.821) = 0.2058$.

   (d) $P(\text{Type I error}) = \alpha = 0.05$; $P(\text{Type II error when } p = 0.6) = 1 - 0.2058 = 0.7942$.

12.20 This exercise is a follow-up to Activity 12.

12.21

X	n	$\hat{p}$	z	P-value
14	50	.28	−.752	.2261
98	350	.28	−1.998	.0233
140	500	.28	−2.378	.0088

Although Tonya, Frank, and Sarah all recorded the same sample proportion, $\hat{p} = .28$, the P-values were all quite different. Conclude: For a given sample proportion, the larger the sample size, the smaller the P-value.

12.22 (a) $\hat{p}_1 = \frac{6}{53} \doteq 0.1132$ and $\hat{p}_2 = \frac{45}{108} \doteq 0.4167$. (b) SE $= \sqrt{\frac{\hat{p}_1(1 - \hat{p}_1)}{53} + \frac{\hat{p}_2(1 - \hat{p}_2)}{108}} \doteq 0.06438$, so the 95% confidence interval is $(0.1132 - 0.4167) \pm (1.96)(0.06438) = -0.4297$ to -0.1773. As the problem notes, this interval should at least apply to the two groups about whom we have information: skaters (with and without wrist guards) with severe enough injuries to go to the emergency room. (c) $45/206 \doteq 21.8\%$ did not respond. If those who did not respond were different from those who did, then not having them represented in our sample makes our conclusions suspect.

12.23 $\hat{p}_1 = \frac{54}{72} = 0.75$ and $\hat{p}_2 = \frac{10}{17} \doteq 0.5882$. Then SE $= \sqrt{\frac{\hat{p}_1(1 - \hat{p}_1)}{72} + \frac{\hat{p}_2(1 - \hat{p}_2)}{17}} \doteq 0.1298$, so the 90% confidence interval is $(0.75 - 0.5882) \pm (1.645)(0.1298) = -0.0518$ to 0.3753. These methods can be used because the populations of mice are certainly more than 10 times as large as the samples, and the counts of successes and failures are more than 5 in both samples.

12.24 The population-to-sample ratios are large enough, and all relevant counts are at least 10. $\hat{p}_1 = 0.3895$, $\hat{p}_2 = 0.3261$, and the interval is -0.0208 to 0.14764.

12.25 To test H_0: $p_1 = p_2$ versus H_a: $p_1 > p_2$, we find $\hat{p}_1 = \frac{223}{33809} \doteq 0.006596$, $\hat{p}_2 = \frac{7}{1541} \doteq 0.004543$ and $\hat{p} = \frac{223 + 7}{33809 + 1541} \doteq 0.006506$. Then SE $= \sqrt{\hat{p}(1 - \hat{p})(\frac{1}{33809} + \frac{1}{1541})} \doteq 0.002094$, so $z = (\hat{p}_1 - \hat{p}_2)/$SE $\doteq 0.98$. This gives $P = 0.1635$, so we don't have enough evidence to conclude that the death rates are different.

12.26 (a) Diagram below. (b) To test H_0: $p_1 = p_2$ versus H_a: $p_1 \neq p_2$, we find $\hat{p}_1 = \frac{206}{1649} \doteq 0.1249$, $\hat{p}_2 = \frac{157}{1650} \doteq 0.09515$, and $\hat{p} = \frac{206 + 157}{1649 + 1650} \doteq 0.11003$. Then SE $= \sqrt{\hat{p}(1 - \hat{p})(\frac{1}{1649} + \frac{1}{1650})} \doteq 0.010897$, so $z = (\hat{p}_1 - \hat{p}_2)/$SE $\doteq 2.73$. This gives $P = 0.0064$, which is very strong evidence that the proportions are different. (c) With the same hypotheses (for the proportions of deaths), this time we find $\hat{p}_1 = \frac{182}{1649} \doteq 0.1104$, $\hat{p}_2 = \frac{185}{1650} \doteq 0.1121$, and $\hat{p} = \frac{182 + 185}{1649 + 1650} \doteq 0.11125$. Then SE $\doteq 0.010949$, so $z \doteq -0.16$; this gives $P = 0.8728$—virtually no evidence of a difference.

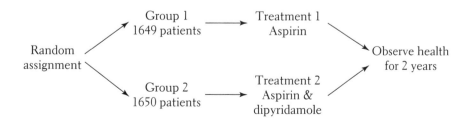

12.27 For home computer access, we test H_0: $p_1 = p_2$ versus H_a: $p_1 \neq p_2$ and find $\hat{p}_1 = \frac{86}{131} \doteq 0.6565$, $\hat{p}_2 = \frac{1173}{1916} \doteq 0.6122$, and $\hat{p} = \frac{86 + 1173}{131 + 1916} \doteq 0.6150$. Then SE $= \sqrt{\hat{p}(1 - \hat{p})(\frac{1}{131} + \frac{1}{1916})} \doteq 0.04394$, so $z = (\hat{p}_1 - \hat{p}_2)/$SE $\doteq 1.01$. This gives $P = 0.3124$—little evidence that the proportions are different. With the same hypotheses for the proportions with PC access at work, we find

$\hat{p}_1 = \frac{100}{131} \doteq 0.7634$, $\hat{p}_2 = \frac{1132}{1916} \doteq 0.5908$, and $\hat{p} = \frac{100 + 1132}{131 + 1916} \doteq 0.6019$. Then SE $\doteq 0.044207$, so $z \doteq 3.90$. This gives $P < 0.0004$, so we have very strong evidence of a difference (specifically, that higher-income blacks have more access at work).

12.28 (a) For eventual contact, we test H_0: $p_1 = p_2$ versus H_a: $p_1 < p_2$ and find $\hat{p}_1 = \frac{58}{100} = 0.58$, $\hat{p}_2 = \frac{200}{291} \doteq 0.6873$, and $\hat{p} = \frac{58 + 200}{100 + 291} \doteq 0.6598$. Then SE $= \sqrt{\hat{p}(1 - \hat{p})(\frac{1}{100} + \frac{1}{291})} \doteq 0.05492$, so $z = (\hat{p}_1 - \hat{p}_2)/\text{SE} \doteq -1.95$. This gives $P = 0.0254$—fairly strong evidence that the proportions and different.

(b) With the same hypotheses for the proportions completing the survey, we find $\hat{p}_1 = \frac{33}{100} = 0.33$, $\hat{p}_2 = \frac{134}{291} \doteq 0.4605$, and $\hat{p} = \frac{33 + 134}{100 + 291} \doteq 0.4271$. Then SE $\doteq 0.05734$, so $z \doteq -2.28$. This gives $P = 0.0113$, so we have strong evidence of a difference.

(c) For survey completion, the standard error for a confidence interval is SE $= \sqrt{\frac{\hat{p}_1(1 - \hat{p}_1)}{100} + \frac{\hat{p}_2(1 - \hat{p}_2)}{291}} \doteq 0.05536$, so, for example, a 95% confidence interval for the difference $p_1 - p_2$ is $(0.33 - 0.4605) \pm (1.96)(0.05536) = -0.2390$ to -0.0220. Although the difference for "eventual contact" was not significant, we might examine the difference anyway: The confidence interval standard error is SE $\doteq 0.05634$, so a 95% confidence interval for the difference $p_1 - p_2$ is $(0.58 - 0.6873) \pm (1.96)(0.05634) = -0.2177$ to 0.0031. This indicates that the differences could be very small, but they might be substantially large, and so have the potential to reduce nonresponse rates in surveys. At the very least, further study is warranted.

12.29 (a) H_0: $p_1 = p_2$ vs. H_a: $p_1 > p_2$; the populations are much larger than the samples, and 17 (the smallest count) is greater than 5.

(b) $\hat{p} = 0.0632$, $z = 2.926$, and $P = 0.0017$—the difference is statistically significant.

(c) Neither the subjects nor the researchers who had contact with them knew which subjects were getting which drug—if anyone had known, they might have confounded the outcome by letting their expectations or biases affect the results.

12.30 (a) H_0: $p_1 = p_2$ vs. H_a: $p_1 \neq p_2$; $P = 0.0028$—the difference is statistically significant. (b) 0.1172 to 0.3919.

12.31 H_0: $p_1 = p_2$ vs. H_a: $p_1 \neq p_2$; $P = 0.9805$—insufficient evidence to reject H_0.

12.32 (a) H_0: $p_1 = p_2$ vs. H_a: $p_1 > p_2$; $P = 0.0335$—reject H_0 (at the 5% level). (b) -0.0053 to 0.2336.

12.33 The population-to-sample ratio is large enough, and the smallest count is 10—twice as big as it needs to be to allow the z procedures. Fatal heart attacks: $z = -2.67$, $P = 0.0076$. Non-fatal heart attacks: $z = -4.58$, $P < 0.000005$. Strokes: $z = 1.43$, $P = 0.1525$. The proportions for both kinds of heart attacks were significantly different; the stroke proportions were not.

12.34 (a) -9.91% to -4.09%. Since 0 is not in this interval, we would reject H_0: $p_1 = p_2$ at the 1% level. (b) -0.7944 to -0.4056. Since 0 is not in this interval, we would reject H_0: $\mu_1 = \mu_2$ at the 1% level.

12.35 (a) $\hat{p} = \frac{5690}{12,931} \doteq 0.4400$ and $\text{SE}_{\hat{p}} = \sqrt{(0.4400)(0.5600)/12,931} \doteq 0.004365$, so the 95% confidence interval is $0.4400 \pm (1.96)(0.004365) \doteq 0.4315$ to 0.4486.

(b) $\hat{p} \approx 0.4400$, $\hat{p}_2 \approx \frac{1051}{3285} = 0.3200$, and SE ≈ 0.0092; the 95% confidence interval is $(0.4400 - 0.3200) \pm (1.96)(0.0092)$, or 0.102 to 0.138.

(c) In a cluster of cars, where one driver's behavior might affect the others, we do not have *independence*—one of the important properties of a random sample.

12.36 (a) We must make sure that we draw the samples from a representative cross-section of all Illinois high-school freshmen and seniors. A multistage/cluster sampling procedure could be used in order to ensure that all geographical areas of the state are equally likely to contain sampled individuals.

(b) $\hat{p} = .02025$, and since $n\hat{p} = 34$ and $n(1 - \hat{p}) = 1645$ are both greater than 10, the confidence interval based on z can be used. The 95% confidence interval for p is $.02025 \pm (1.96)\sqrt{((.02025)(.97975)/1679)} = .02025 \pm 0.00674$, or 0.01351 to 0.02699.

(c) Letting $p_1 =$ the proportion of freshmen who have used steroids and $p_2 =$ the proportion of seniors who have used steroids, we test $H_0\colon p_1 = p_2$ vs. $H_a\colon p_1 \neq p_2$. From the data, $\hat{p}_1 = .02025, \hat{p}_2 = .01057$, and $\hat{p} = .01905$. $n_1\hat{p}, n_1(1 - \hat{p}), n_2\hat{p}$, and $n_2(1 - \hat{p})$ are all greater than 5, so a normal approximation can be used. Test statistic $z = (.02025 - .01757)/\sqrt{((.01905)(.98095)(1/1679 + 1/1366)} = 0.5382$, and the P-value $= 0.59$. There is no reason to reject H_0; the difference between p_1, p_2 is not significant.

12.37 (a) $\hat{p} \doteq 0.1486$ and $SE_{\hat{p}} = \sqrt{(0.1486)(0.8514)/148} \doteq 0.02923$, so the 95% confidence interval is $0.1486 \pm (1.96)(0.02923) \doteq 0.0913$ to 0.2059.

(b) $n = \left(\frac{1.96}{0.04}\right)^2(0.1486)(0.8514) \doteq 303.7$—use $n = 304$. (We should not use $p^* = 0.5$ here since we have evidence that the true value of p is not in the range 0.3 to 0.7.)

(c) Aside from the 45% nonresponse rate, the sample comes from a limited area in Indiana, focuses on only one kind of business, and leaves out any businesses not in the Yellow Pages (there might be a few of these; perhaps they are more likely to fail). It is more realistic to believe that this describes businesses that match the above profile; it *might* generalize to food-and-drink establishments elsewhere, but probably not to hardware stores and other types of business.

12.38 $H_0\colon p_1 = p_2$ vs. $H_a\colon p_1 \neq p_2$; $P = 0.6981$—insufficient evidence to reject H_0.

12.39 (a) $\hat{p}_m = 0.1415, \hat{p}_w = 0.1667$; $P = 0.6981$. (b) $z = 2.12$, $P = 0.0336$. (c) From (a): -0.1056 to 0.1559. From (b): 0.001278 to 0.049036. The larger samples make the margin of error (and thus the length of the confidence interval) smaller.

12.40 For testing $H_0\colon p = 1/3$ versus $H_a\colon p > 1/3$, we have $\hat{p} \doteq 0.3786$, and the test statistic is $z = (0.3786 - 1/3)/\sqrt{\frac{(1/3)(2/3)}{803}} \doteq 2.72$. This gives $P = 0.0033$—very strong evidence that more than one-third of this group never use condoms.

12.41 (a) 0.2465 to 0.3359—since 0 is not in this interval, we would reject $H_0\colon p_1 = p_2$ at the 1% level (in fact, P is practically 0). (b) No: $t = -0.8658$, which gives a P-value close to 0.4.

12.42 No—the data is not based on an SRS, and thus the z procedures are not reliable in this case. In particular, a voluntary response sample is typically biased.

12.43 $\hat{p} = \frac{427}{3160} \doteq 0.1351$, and $SE_{\hat{p}} = \sqrt{\hat{p}(1 - \hat{p})/3160} \doteq 0.006081$, so the 99% confidence interval is $0.1351 \pm (2.576)(0.006081) = 0.1194$ to 0.1508.

12.44 To test $H_0\colon p_1 = p_2$ vs. $H_a\colon p_1 < p_2$, we find $\hat{p}_1 = \frac{40}{244} \doteq 0.1639$, $\hat{p}_2 = \frac{87}{245} \doteq 0.3551$, and the pooled value $\hat{p} = \frac{40 + 87}{244 + 245} \doteq 0.2597$. Then $SE = \sqrt{\hat{p}(1 - \hat{p})(\frac{1}{244} + \frac{1}{245})} \doteq 0.03966$, so $z = (\hat{p}_1 - \hat{p}_2)/SE \doteq -4.82$. This gives a tiny P-value (7.2×10^{-7}), so we conclude that bupropion increases the success rate.

12.45 (a) $p_0 = \frac{143,611}{181,535} \doteq 0.7911$. (b) $\hat{p} = \frac{339}{870} \doteq 0.3897$, $\sigma_{\hat{p}} \doteq 0.0138$, and $z = (\hat{p} - p_0)/\sigma_{\hat{p}} \doteq -29.1$, so $P \doteq 0$ (regardless of whether H_a is $p < p_0$ or $p \neq p_0$). This is very strong evidence against H_0; we conclude that Mexican Americans are underrepresented on juries. (c) $\hat{p}_1 = \frac{339}{870} \doteq 0.3897$, while $\hat{p}_2 = \frac{143,611-339}{181,535-870} \doteq 0.7930$. Then $\hat{p} \doteq 0.7911$ (the value of p_0 from part (a)), $s_p = 0.0138$, and $z \doteq -29.2$—and again, we have a tiny P-value and reject H_0.

13

Inference for Tables

13.1 (a) (i) $0.20 < P < 0.25$. (ii) $P = 0.235$. (b) (i) $0.02 < P < 0.025$. (ii) $P = 0.0204$. (c) (i) $P > 0.25$. (ii) $P = 0.3172$.

13.2 H_0: The marital-status distribution of 25- to 29-year-old U.S. males is the same as that of the population as a whole. H_a: The marital-status distribution of 25- to 29-year-old U.S. males is different from that of the population as a whole. Expected counts: 140.5, 281.5, 32, 46. $X^2 = 161.77$, df = 3. P-value = $7.6 \times 10^{-35} \approx 0.0000$. Reject H_0. The two distributions are different.

13.3 H_0: The genetic model is valid (the different colors occur in the stated ratio of 1:2:1). H_a: The genetic model is not valid. Expected counts: 21 GG, 42 Gg, 21 gg. $X^2 = 5.43$, df = 2. P-value = $P(X_2^2 > 5.43) = 0.0662$. There is no compelling reason to reject H_0 (though the P-value **is** a little on the low side).

13.4 H_0: The ethnicity distribution of the Ph.D. degree in 1994 is the same as it was in 1981. H_a: The ethnicity distribution of the Ph.D. degree in 1994 is different from the distribution in 1981. Expected counts = $300 \times (1981 \text{ percents}) = 237, 12, 4, 8, 1, 38$. $X^2 = 61.98$, df = 5. P-value = $P(X_5^2 > 61.98) = 4.734 \times 10^{-12} \approx 0.0000$. We reject H_0 and conclude that the ethnicity distribution of the Ph.D. degree has changed from 1981 to 1994. (b) The greatest change is that many more nonresident aliens than expected received the Ph.D. degree in 1994 over the 1981 figures. To a lesser extent, a smaller proportion of white, non-Hispanics received the Ph.D. degree in 1994.

13.5 Use a χ^2 goodness of fit test. (b) Use a one-proportion z test. (c) You can construct the interval; however, your ability to generalize may be limited by the fact that your sample of bags is not an SRS. M&M's may be packaged by weight rather than count.

13.6 H_0: The age-group distribution in 1996 is the same as the 1980 distribution. H_a: The age-group distribution in 1996 is different from the 1980 distribution. One simulation produced observed counts: 37, 35, 15, 13. The expected counts: 41.39, 27.68, 19.64, and 11.28 are stored in list L_4, and the difference terms $(O - E)^2/E$ are assigned to L_5.

```
HOW MANY TRIALS
N=?100

OBSERVED COUNTS
ARE IN L3
                Done
■
```

```
L1      L2     L3      3
64      0      37
11      0      35
10      0      15
32      0      13
44      0      ------
93      1
13      0
L3(1)=37
```

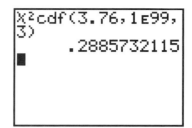

Note that none of the difference terms is very large. The test statistic is $X^2 = 3.76$, df $= 3$, and the P-value $= .2886$. There is insufficient evidence to conclude that the distributions are different. The results of this simulation differ from those in the text; the reason may be due to different sample sizes or simply chance.

13.7 (a) $H_0 : p_0 < p_1 = p_2 = \ldots = p_9 = 0.1$ vs. H_a: At least one of the p's is not equal to 0.1. (b) Using randInt $(0, 9, 200) \rightarrow L_4$, we obtained these counts for digits 0 to 9: 19, 17, 23, 22, 19, 20, 25, 12, 27, 16. (e) $X^2 = 8.9$, df $= 9$, P-value $= .447$. There is no evidence that the sample data were generated from a distribution that is different from the uniform distribution.

13.8 Answers will vary. You should be surprised if you get a significant P-value. H_0: The die is fair ($p_1 = p_2 = \ldots = p_6 = 1/6$). H_a: The die is not fair. Use the command randInt $(1, 6, 300) \rightarrow L_1$ to simulate rolling a fair die 300 times. In our simulation, we obtained the following frequency distribution:

Side	1	2	3	4	5	6
Freq.	57	46	55	54	45	43

The expected counts under H_0: $(300)(1/6) = 50$ for each side. The test statistic is $X^2 = .98 + .32 + .5 + .32 + .5 + .98 = 3.6$, and the degrees of freedom are $n - 1 = 5$. The P-value is $P(X_5^2 > 3.6) = .608$. Since the P-value is large, we fail to reject H_0. There is no evidence that the die is not fair.

13.9 (a)

Outcome	H	T
Frequency	78	122
Expected	100	100

H_0: The distribution of heads and tails from spinning a 1982 penny shows equally likely outcomes. H_a: Heads and tails are not equally likely. df $= 1$ and $X^2 = 4.84 + 4.84 = 9.68$. The P-value is $P(X_1^2 > 9.68) = .00186$. Reject H_0 and conclude that spinning a 1982 penny does not produce equally likely results.

(b) We will test H_0: $p = 0.5$ vs. H_a: $p \neq 0.5$, where p = probability of getting tails when the coin is spun. Since $np_0 = n(1 - p_0) = 100 > 10$, the z-test for a single proportion may be used. Test statistic $z = 3.111$, P-value of test $= 0.00186$. Reject H_0; heads and tails are clearly not equally likely.

(c) The p-values are identical.

13.10 Let $p_1, p_2, \ldots, p_6$ denote the probability of getting a 1, 2, 3, $\ldots$, 6. If the die is fair, then $p_1 = p_2 = \ldots = p_6$. H_0: $p_1 = p_2 = \ldots = p_6$ (die is fair). H_a: The die is "loaded"/unfair. The observed counts for sides $1 - 6$ are: 26, 36, 39, 30, 38, 32. The expected counts are $(200)(1/6) = 33.33$ for each side. df $= 6 - 1 = 5$, and $X^2 = 1.612 + 0.214 + 0.965 + 0.333 + 0.654 + 0.053 = 3.831$. The P-value is $P(X_5^2 > 3.831) = .574$. Since this P-value is rather large, we fail to reject H_0, and conclude that there is no evidence that the die is "loaded."

13.11 The observed and expected values are:

Flavor	Grape	Lemon	Lime	Orange	Strawberry
Observed	530	470	420	610	585
Expected	523	523	523	523	523

H_0: Trix flavors are uniformly distributed. H_a: The flavors are not uniformly distributed. df = 5 − 1 = 4, and X^2 = .09369 + 5.3709 + 20.285 + 14.472 + 7.3499 = 47.57. $P(X_4^2 > 47.57) = 1.16 \times 10^{-9} \doteq .0000$. Reject H_0 and conclude that either the Trix flavors are *not* uniformly distributed, or our box of Trix is not a *random* sample.

13.12 Answers will vary.

13.13 Since the wheel is divided into four equal parts, if it is in balance, then the four outcomes should occur with approximately equal frequency. Here are the observed and expected values:

Parts	I	II	III	IV
Observed	95	105	135	165
Expected	125	125	125	125

H_0: The wheel is balanced (the four outcomes are uniformly distributed). H_a: The wheel is not balanced. df = 3 and X^2 = 7.2 + 3.2 + 0.8 + 12.8 = 24. The P-value is $P(X_3^2 > 24) = 2.5 \times 10^{-5} = .000025$. Reject H_0 and conclude that the wheel is not balanced. Since "Part IV: Win nothing" shows the greatest deviation from the expected result, there may be reason to suspect that the carnival game operator may have tampered with the wheel to make it harder to win.

13.14 (a)

Treatment	Successes	Failures
Nicotine patch	40	204
Drug	74	170
Patch plus drug	87	158
Placebo	25	135

(b) The success rates are $\frac{40}{244} \doteq 0.1639$, $\frac{74}{244} \doteq 0.3033$, $\frac{87}{245} \doteq 0.3551$, and $\frac{25}{160} = 0.15625$.
(c)

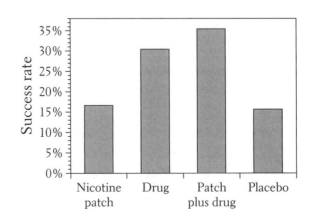

(d) The success rate (proportion of those who quit) is the same for all four treatments.

(e) Using the formula for expected counts, we obtain the following table:

Treatment	Exp. Succ.	Exp. Fail.
Nicotine patch	61.75	182.25
Drug	61.75	182.25
Patch plus drug	62	183
Placebo	40.49	119.51

(f) A higher percentage than expected of the "patch plus drug" and "drug" subjects successfully quit, and a lower percentage than expected of the other two groups quit. This reflects the fact that the "patch plus drug" and "drug" success rates were considerably higher than the success rates for the other two groups, as seen in (b) and (c).

13.15 (a) r = the number of rows in the table, c = the number of columns in the table. In this case $r = 4$ and $c = 2$.

(b) Females: 20.9% HSC-HM, 10.4% HSC-LM, 31.3% LSC-HM, 37.3% LSC-LM. Males: 46.3% HSC-HM, 26.9% HSC-LM, 7.5% LSC-HM, 19.4% LSC-LM.

(c)

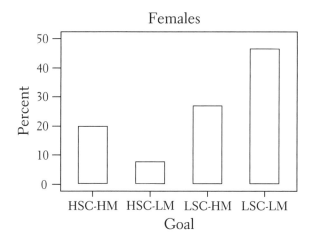
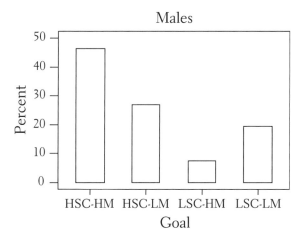

(d)

Goal	Exp. Counts (Females)	Exp. Counts (Males)
HSC-HM	22.5	22.5
HSC-LM	12.5	12.5
LSC-HM	13	13
LSC-LM	19	19

(e) Males are more likely to fall into the HSC categories (HSC-HM and HSC-LM) than their expected counts would predict. Likewise, females are more likely to fall into the LSC categories (LSC-HM and LSC-LM). It appears that males and females have distinctly different goals when they play sports.

13.16 (a) Reading row by row, the components are 7.662, 2.596, 2.430, 0.823, 10.076, 3.414, 5.928, 2.008. $X^2 = 34.937$ (df = 3).

(b) According to Table E, P-value = $P(X_3^2 > 34.937) < 0.0005$. Since the largest critical value for df = 3 is 17.73, the P-value is actually quite a bit smaller than 0.0005. A P-value of this size indicates that it is extremely unlikely that such a result occurred due to chance; it represents very strong evidence against H_0.

(c) 10.076, corresponding to the "patch plus drug/success" category, is the largest contributor. This is not surprising because, according to Exercise 13.14(f), the "patch plus drug" group contains a higher than expected number of successful quitters.

(d) Treatment is strongly associated with success; specifically, the drug, or the patch together with the drug, seem to be most effective.

(e) See displays below.

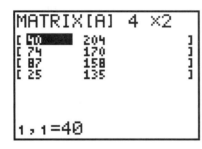
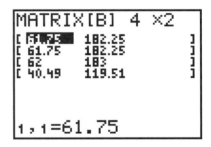

13.17 (a) Reading row by row, the components are 3.211, 3.211, 2.420, 2.420, 4.923, 4.923, 1.895, 1.895. $X^2 = 24.898$ (df = 3).

(b) From Table E, $P(X_3^2 > 24.898) < 0.0005$. A P-value of this size indicates that it is extremely unlikely that such a result occurred due to chance; it represents very strong evidence against H_0.

(c) The terms corresponding to HSC-HM and LSC-HM (for both sexes) provide the largest contributions to X^2. This reflects the fact that males are more likely to have "winning" (social comparison) as a goal, while females are more concerned with "mastery."

(d) The terms and results are identical. The P-value of 0.000 in the MINITAB output reflects the fact that the true P-value in part (b) was actually considerably smaller than 0.0005.

13.18 Various graphs can be made; one possibility is shown below. For the null hypothesis "There is no relationship between race and opinions about schools," we find $X^2 = 22.426$ (df = 8) and $P = 0.004$ (Minitab output below). We have evidence that there *is* a relationship; specifically, blacks are less likely, and Hispanics more likely, to consider schools "excellent," while Hispanics and whites differ in percentage considering schools "good" (whites are higher) and percentage who "don't know" (Hispanics are higher). Also, a higher percentage of blacks rated schools as "fair."

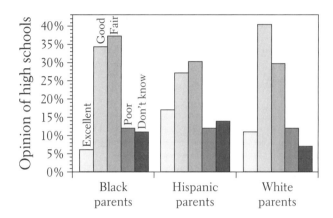

Minitab output

	Black	Hispanic	White	Total
1	12	34	22	68
	22.70	22.70	22.59	
2	69	55	81	205
	68.45	68.45	68.11	
3	75	61	60	196
	65.44	65.44	65.12	
4	24	24	24	72
	24.04	24.04	23.92	
5	22	28	14	64
	21.37	21.37	21.26	
Total	202	202	201	605

ChiSq = 5.047 + 5.620 + 0.015 +
        0.004 + 2.642 + 2.441 +
        1.396 + 0.301 + 0.402 +
        0.000 + 0.000 + 0.000 +
        0.019 + 2.058 + 2.481 = 22.426

df = 8, p = 0.004

13.19 (a) $r = 2$, $c = 3$.

(b) 55.0%, 74.7%, and 37.5%. Some (but not too much) time spent in extracurricular activities seems to be beneficial.

(c)

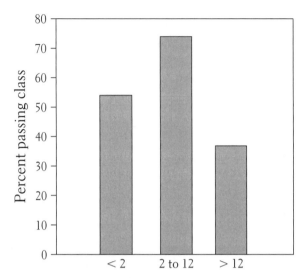

(d) H_0: There is no association between amount of time spent on extracurricular activities and grades earned in the course.

H_a: There is an association.

(e)

	<2	2–12	>12
C or better	13.78	62.71	5.51
D or F	6.22	28.29	2.49

(f) The first and last columns have lower numbers than we expect in the "passing" row (and higher numbers in the "failing" row), while the middle column has this reversed—more passed than we would have expected if the proportions were all equal.

13.20 (a) 3×2. (b) 22.5%, 18.6%, and 13.9%. A student's likelihood of smoking increases when one parent smokes, and increases even more when both smoke. (c) See Exercise 4.53. (d) The null hypothesis says that parents' smoking habits have no effect on their children. (e) Below. (f) In column 1, row 1, the expected count is much smaller than the actual count; meanwhile, the actual count is lower than expected in the lower left. This agrees with what we observed before: Children of non-smokers are less likely to smoke.

	Student smokes	Student does not smoke
Both parents smoke	332.49	1447.51
One parent smokes	418.22	1820.78
Neither parent smokes	253.29	1102.71

13.21 (a) Missing entries in table of expected counts (row by row): 62.71, 5.51, 6.22. Missing entries in components of X^2: 0.447, 0.991.

(b) df = 2, $P = 0.0313$. Rejecting H_0 means that we conclude that there is a relationship between hours spent in extracurricular activities and performance in the course.

(c) The highest contribution comes from column 3, row 2 (" > 12 hours of extracurricular activities, D or F in the course"). Too much time spent on these activities seems to hurt academic performance.

(d) No—this study demonstrates association, not causation. Certain types of students may tend to spend a moderate amount of time in extracurricular activities and also work hard on their classes—one does not necessarily cause the other.

13.22 (a) H_0: There is no association between smoking by parents and smoking by high school students. H_a: There is an association between smoking by parents and smoking by high school students. Expected counts: given in Exercise 13.20(e). $X^2 = 37.566$, df = 2. The P-value is approximately 7×10^{-9}—essentially 0.

By rejecting H_0, we conclude that there is a relationship between parents' smoking habits and those of their children.

(b) The highest contributions come from Cl R1 ("both parents smoke, student smokes") and Cl R3 ("neither parent smokes, student smokes"). When both parents smoke, their student is much more likely to smoke; when neither parent smokes, their student is unlikely to smoke.

(c) No—this study demonstrates association, not causation. There may be other factors (heredity or environment, for example) that cause *both* students and parent(s) to smoke.

13.23 The proportions with access to a car were $\frac{122}{356} \doteq 0.3427$, $\frac{138}{318} \doteq 0.4340$, and $\frac{213}{555} \doteq 0.3838$ (graph below). To test the null hypothesis (sleep patterns do not affect car access), we find $X^2 = 5.915$ (df = 2), and $P = 0.052$. While the data suggest that owls are more likely to have access to a car, we find that the evidence is not quite significant (at the $\alpha = 0.05$ level).

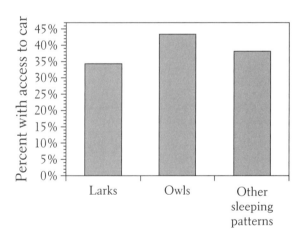

Minitab output

```
              Yes         No      Total
    1         122        234        356
            137.01     218.99

    2         138        180        318
            122.39     195.61

    3         213        342        555
            213.60     341.40

Total         473        756       1229

ChiSq =  1.645 +  1.029 +
         1.992 +  1.246 +
         0.002 +  0.001  = 5.915
df =  2, p = 0.052
```

13.24 (a) H_0: $p_1 = p_2$ vs. H_a: $p_1 \neq p_2$. $z = -0.5675$ and $P = 0.5704$. (b) Table of expected counts is given below. $X^2 = 0.322$, which equals z^2. With 1 df, Table E tells us that $P > 0.25$; a statistical calculator gives $P = 0.5704$. (c) Gastric freezing is not significantly more (or less) effective than a placebo treatment.

	Improved	Did not improve
Gastric freezing	28	54
	29.73	52.28
Placebo	30	48
	28.27	49.72

13.25 (a)

	Cardiac event?		
Group	Yes	No	TOTAL
Stress management	3	30	33
Exercise	7	27	34
Usual care	12	28	40
TOTAL	22	85	107

(b) Success rates (% of "No"s): 90.91%, 79.41%, 70%.

(c)

	Cardiac event?	
Group	Exp. Yes	Exp. No
Stress management	6.785	26.215
Exercise	6.991	27.009
Usual care	8.224	31.776

All expected cell counts exceed 5, so the chi-square test can be used.

(d) $X^2 = 4.84$ (df = 2), P-value = 0.0889. Though the success rate for the stress management group is slightly higher than for the other two groups, there does **not** appear to be a significant difference among the success rates.

13.26 (a) The proportions in favor are $\frac{58}{116} = 0.5$, $\frac{84}{213} \doteq 0.3944$, $\frac{169}{463} \doteq 0.3650$, $\frac{98}{233} \doteq 0.4206$, and $\frac{77}{176} = 0.4375$. Those who did not complete high school and those with a college or graduate degree appear to be more likely to favor a ban. (b) With $X^2 = 8.525$ and 4 degrees of freedom, we have $P = 0.075$, so we cannot reject the null hypothesis; we do not have enough evidence to conclude that the proportion favoring a handgun ban varies significantly with level of education.

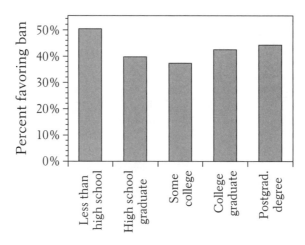

Minitab output

	Yes	No	Total
1	58 46.94	58 69.06	116
2	84 86.19	129 126.81	213
3	169 187.36	294 275.64	463
4	98 94.29	135 138.71	233
5	77 71.22	99 104.78	176
Total	486	715	1201

```
ChiSq = 2.605 + 1.771 +
        0.056 + 0.038 +
        1.799 + 1.223 +
        0.146 + 0.099 +
        0.469 + 0.319 = 8.525

df = 4, p = 0.075
```

13.27 H_0: all proportions are equal vs. H_a: some proportions are different. Table below. $X^2 = 10.619$ with 2 df; and $P = 0.0049$—good evidence against H_0, so we conclude that contact method makes a difference in response.

	Yes	No
Phone	168	632
One-on-one	200	600
Anonymous	224	576

13.28 (a) 7.01%, 14.02%, and 13.05%. (b) and (c) Table below—actual counts above, expected counts below. Expected counts are all much bigger than 5, so the chi-square test is safe. H_0: there is no relationship between worker class and race vs. H_a: there is some relationship. (d) df = 2; $P < 0.0005$ (basically 0). (e) Black female child-care workers are more likely to work in non-household or preschool positions.

	Black	Other
Household	172 242.36	2283 2212.64
Nonhousehold	167 117.58	1024 1073.42
Teachers	86 65.06	573 593.94

13.29 (a) H_0: $p_1 = p_2$, where p_1 and p_2 are the proportions of women customers in each city. $\hat{p}_1 = 0.8423, \hat{p}_2 = 0.6881, z = 3.9159$, and $P = 0.00009$. (b) $X^2 = 15.334$, which equals z^2. With df = 1, Table E tells us that $P < 0.0005$; a statistical calculator gives $P = 0.00009$. (c) 0.0774 to 0.2311.

13.30 4 degrees of freedom; $P > 0.25$ (in fact, $P = 0.4121$). There is not enough evidence to reject H_0 at any reasonable level of significance; the difference in the two income distributions is not statistically significant.

13.31 The observed frequencies of scores in this sample, their marginal percents, and the expected numbers were:

Score	5	4	3	2	1
Frequency	167	158	101	79	30
Percent	31.2	29.5	18.9	14.8	5.6
Expected	81.855	117.7	132.68	105.93	96.835

H_0: The distribution of scores in this sample is the same as the distribution of scores for all students who took this inaugural exam. H_a: The distribution of scores in this sample is different from the national results. The degrees of freedom are $n - 1 = 4$, and the chi-square statistic is $X^2 = 88.57 + 13.80 + 7.56 + 6.85 + 46.13 = 162.9$. The P-value is $P(X_4^2 > 162.9) = 3.49 \times 10^{-34} = .0000$. Reject H_0 and conclude that the distribution of AP Statistics exam scores in this sample is different from the national distribution.

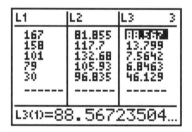

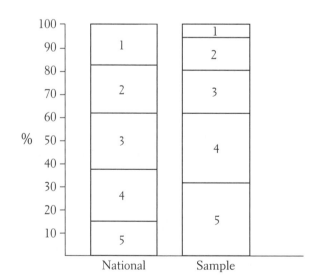

Postscript: As soon as the exam grades were sent to the students and their schools, in July 1997, several AP Statistics teachers who were subscribers to an AP Statistics discussion group on the Internet posted their grades in the spirit of sharing the results with their fellow teachers. While this was of interest to many of the pioneering AP Statistics teachers in their first year teaching this new course, this sample was a voluntary response sample, not a random sample. It should come as no surprise that these self-reported results were weighted toward the higher scores.

13.32 There is no reason to consider one of these variables as explanatory, but a conditional distribution is useful to determine the nature of the association. Each cell in the table below contains a pair of percentages; the first is the column percent, and the second is the row percent. For example, among nonsmokers, 34.5% were nondrinkers; among nondrinkers, 85.4% were nonsmokers. The percentages in the right margin gives the distribution of alcohol consumption (the overall column percent), while the percentages in the bottom margin are the distribution of smoking behavior.

	0 mg	1−15 mg	16+ mg	
0 oz	105	7	11	123
	82.73	17.69	22.59	
	34.5%	10.8%	13.3%	27.2%
	85.4%	5.7%	8.9%	
0.01− 0.10 oz	58	5	13	76
	51.12	10.93	13.96	
	19.1%	7.7%	15.7%	16.8%
	76.3%	6.6%	17.1%	
0.11− 0.99 oz	84	37	42	163
	109.63	23.44	29.93	
	27.6%	56.9%	50.6%	36.1%
	51.5%	22.7%	25.8%	
1.00+ oz	57	16	17	90
	60.53	12.94	16.53	
	18.8%	24.6%	20.5%	19.9%
	63.3%	17.8%	18.9%	
	304	65	83	452
	67.3%	14.4%	18.4%	100%

$X^2 = 42.252$ (df = 6) so $P < 0.0005$; we conclude that alcohol and nicotine consumption are not independent. The chief deviation from independence (based on comparison of expected and actual counts) is that nondrinkers are more likely to be nonsmokers than we might expect, while those drinking 0.11 to 0.99 oz/day are less likely to be nonsmokers than we might expect. One possible graph is below.

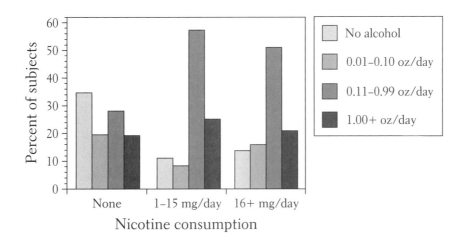

13.33 (a) $H_0: p_1 = p_2$ vs. $H_a: p_1 < p_2$. The z test must be used because the chi-square procedure will not work for a one-sided alternative. (b) $z = -2.8545$ and $P = 0.0022$. Reject H_0; there is strong evidence in favor of H_a.

13.34 (a) Minitab output below. There is strong evidence of a relationship: $X^2 = 11.141$ (df = 2), which means that $0.0025 < P < 0.005$. The second row of the table accounts for most of this; those two cells contribute $3.634 + 4.873 = 8.507$ to the value of X^2.

(b) Minitab output (including the two-way table) below. The relationship is still significant ($X^2 = 10.751$, df = 1, $0.0005 < P < 0.001$). A lower-than-expected number of identical twins had different behavior, and a (slightly) higher-than-expected number of identical twins had the same behavior.

Minitab output

	Ident	Frat	Total
Neither	443	301	744
	426.17	317.83	
One	102	113	215
	123.16	91.84	
Both	45	26	71
	40.67	30.33	
Total	590	440	1030

```
ChiSq = 0.664 + 0.891 +
        3.634 + 4.873 +
        0.461 + 0.618 = 11.141

df = 2, p = 0.004
```

Minitab output

	Ident	Frat	Total
Same	488	327	815
	466.84	348.16	
Diff	102	113	215
	123.16	91.84	
Total	590	440	1030

```
ChiSq = 0.959 + 1.285 +
        3.634 + 4.873 = 10.751

df = 1, p = 0.001
```

13.35 (a) On the facing page. (b) Cold water: $\hat{p}_1 = \frac{16}{27} \doteq 59.3\%$; Neutral: $\hat{p}_2 = \frac{38}{56} \doteq 67.9\%$; Hot water: $\hat{p}_3 = \frac{75}{104} \doteq 72.1\%$. The percentage hatching increases with temperature; the cold water did not prevent hatching, but made it less likely. (c) The differences are not significant: $X^2 = 1.703$, df = 2, and $P = 0.427$.

	Temperature		
	Cold	Neutral	Hot
Hatched	16	38	75
Did not hatch	11	18	29
Total	27	56	104

Minitab output

```
          Cold     Neutral       Hot      Total
  1         16          38        75        129
          18.63       38.63     71.74

  2         11          18        29         58
           8.37       17.37     32.26

Total       27          56       104        187

ChiSq =  0.370 +  0.010 +  0.148 +
         0.823 +  0.023 +  0.329 =  1.703

df = 2,  p = 0.427
```

13.36 We wish to test H_0: The hypothesized model is correct against H_a: The hypothesized model is not correct. The expected number of green-seeded plants according to the model is $(^3/_4)(880) = 660$, while the expected number of yellow-seeded plants is $(^1/_4)(880) = 220$. The value of the chi-square statistic is $X^2 = \frac{(639 - 660)^2}{660} + \frac{(241 - 220)^2}{220} = 0.668 + 2.0045 = 2.6725$. With df = 1, the P-value is $P(X_1^2 > 2.6725) = 0.1021$. There is no reason to doubt the model.

13.37 (a) No: No treatment was imposed.

(b) See the column percents in the table below. Pet owners seem to have better survival rates.

	No Pet	Pet	
Alive	28 33.07 71.8%	50 44.93 94.3%	78 84.8%
Dead	11 5.93 28.2%	3 8.07 5.7%	14 15.2%
	39	53	92

(c) $X^2 = 0.776 + 0.571 + 4.323 + 3.181 = 8.851$ (df = 1), so $0.0025 < P < 0.005$ (in fact, $P = 0.003$).

(d) Provided we believe that there are no confounding or lurking variables, we reject H_0 and conclude that owning a pet improves survival.

(e) We used a χ^2 test. In a z test, we would test H_0: $p_1 = p_2$ vs. H_a: $p_1 < p_2$, where $p_1 =$ the proportion of non-pet owners who survived and $p_2 =$ the proportion of pet owners who survived. For this test, $\hat{p}_1 = .718, \hat{p}_2 = .943, \hat{p} = .848, z = -2.975$, and the P-value $= 0.0015$. As in the χ^2 test, we reject H_0 and conclude that there is a significant difference in the survival rates. The p-value is half that obtained in (c).

13.38 (a) See the Minitab output below for the two-way table. We find $X^2 = 24.243$ with df = 3—a very significant result ($P < 0.0005$). The most effective treatment is "Both"; the largest contributions to the X^2 statistic come from the first and last rows of the first column. Subjects taking a placebo had many more strokes than expected, while those taking both drugs had fewer strokes.

(b) See the Minitab output below for the two-way table. We find $X^2 = 1.418$ with df = 3, which gives $P = 0.701$; the various drug treatments had no significant effect on deaths.

(c) The combination of both drugs is effective at decreasing the risk of stroke, but no drug treatment had a significant impact on death rate.

Minitab output

	Stroke	NoStroke	Total
Placebo	250	1399	1649
	205.81	1443.19	
Aspirin	206	1443	1649
	205.81	1443.19	
Dipyr.	211	1443	1654
	206.44	1447.56	
Both	157	1493	1650
	205.94	1444.06	
Total	824	5778	6602

ChiSq =   9.487 + 1.353 +
          0.000 + 0.000 +
          0.101 + 0.014 +
         11.629 + 1.658 = 24.243

df = 3, p = 0.000

Minitab output

	Death	NoDeath	Total
Placebo	202	1447	1649
	189.08	1459.92	
Aspirin	182	1467	1649
	189.08	1459.92	
Dipyr.	188	1466	1654
	189.65	1464.35	
Both	185	1465	1650
	189.19	1460.81	
Total	757	5845	6602

ChiSq =   0.883 + 0.114 +
          0.265 + 0.034 +
          0.014 + 0.002 +
          0.093 + 0.012 = 1.418

df = 3, p = 0.701

13.39 (a) Yes, the evidence is *very* strong that a higher proportion of men die ($X^2 = 332.205$, df = 1). Possibly many sacrificed themselves out of a sense of chivalry ("women and children first").

(b) For women, $X^2 = 103.767$ (df = 2)—a very significant difference. Over half of the lowest-status women died, but this percentage drops sharply when we look at middle-status women, and it drops again for high-status women.

(c) For men, $X^2 = 34.621$ (df = 2)—another very significant difference (though not quite so strong as the women's value). Men with the highest status had the highest proportion surviving (over one-third). The proportion for low-status men was only about half as big, while middle-class men fared worst (only 12.8% survived).

13.40 Answers vary.

14

Inference for Regression

14.1 (a) See also the solution to Exercise 3.19. The correlation is $r = 0.994$, and linear regression gives $\hat{y} = -3.660 + 1.1969x$. The scatterplot below shows a strong, positive, linear relationship, which is confirmed by r.

(b) β represents how much we can expect the humerus length to increase when femur length increases by 1 cm, b (the estimate of β) is 1.1969, and the estimate of α is $a = -3.660$.

(c) The residuals are -0.8226, -0.3668, 3.0425, -0.9420, and -0.9110; the sum is -0.0001 (but carrying a different number of digits might change this). Squaring and summing the residuals gives 11.79, so that $s = \sqrt{11.79/3} \doteq 1.982$.

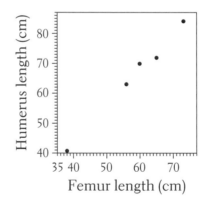

14.2 (a) HEIGHT $= 71.950 + 0.38333$(AGE). The intercept is 71.950 and the slope is 0.38333. (b) The estimate for α is the intercept of the least-squares line, that is, 71.950. The estimate for β is the slope of the least-squares line, that is, 0.38333. (c) The residuals are 0.25012, -0.34984, -0.49983, 0.35018, 0.20019, 0.0502. The formula for s yields $s = \sqrt{\dfrac{.6}{6-2}} = \sqrt{.15} = 0.3873$.

14.3 (a) HEIGHT $= 11.547 + 0.84042$(ARMSPAN). (b) The least-squares line is an appropriate model for the data because the residual plot shows no obvious pattern. (c) $a = 11.547$ estimates the true intercept, α; $b = 0.84042$ estimates the true slope, β. (d) $s = 1.6128$ estimates σ.

14.4 (a) See Exercise 3.71 for scatterplot. $r = 0.9990$ and the equation of the least-squares line is $\hat{y} = 1.766 + 0.080284x$. The scatterplot shows a strong linear relationship, which is confirmed by r.

(b) The residuals are .0107, −.0012, −.001, −.0109, −.0093, .00322, and .00892; the sum is −.00051 (essentially 0).

(c) $a = 1.766$ is the estimate of α; $b = 0.080284$ is the estimate of β;

$$s = \sqrt{\frac{.0004112}{7-2}} = \sqrt{.0000822} = .0091 \text{ is the estimate of } \sigma.$$

14.5 Answers will vary.

14.6 (a) $\hat{y} = -3.6596 + 1.1969x$. (b) $t = b/\text{SE}_b = 1.1969/0.0751 = 15.9374$. (c) df = 3; since $t > 12.92$, we know that $P < 0.0005$. (d) There is very strong evidence that $\beta > 0$, that is, that the line is useful for predicting the length of the humerus given the length of the femur. (e) For df = 3, the critical value for a 99% confidence interval is $t^* = 5.841$. The interval is $1.1969 \pm (5.841)(0.0751)$ or 1.1969 ± 0.439, that is, 0.7579 to 1.6359.

14.7 (a) $H_0: \beta = 0$ (there is no association between number of jet skis in use and number of fatalities). $H_a: \beta > 0$ (there is a positive association between number of jet skis in use and number of fatalities).

(b) The conditions are satisfied except for having independent observations. We will proceed with caution.

(c) LinRegTTest (TI-83) reports that $t = 7.26$ with df = 8. The P-value is 0.000. With the earlier caveat, there is sufficient evidence to reject H_0 and conclude that there is an association between year and number of fatalities. As the number of jet skis in use increases, the number of fatalities increases.

(d) The confidence interval takes the form $b \pm t^*\text{SE}_b$. With $t^* = 2.8214$, and $\text{SE}_b = .00000913$, the 98% confidence interval is approximately (0.00004024, 0.00009176).

14.8 (a) Regression of deaths on wine consumption gives $b = -22.969$, $\text{SE}_b = 3.357$, and $t = -6.46$. With df = 17, we see that $P < 0.0005$, so we have strong evidence that $\beta < 0$ and hence that the correlation is negative.

(b) For a 95% confidence interval with df = 17, $t^* = 2.110$. The 95% confidence interval for β is $-22.969 \pm (2.110)(3.357)$ or -22.969 ± 7.08327, that is, -30.05227 to -15.88573.

14.9 Regression of fuel consumption on speed gives $b = -0.01466$, $\text{SE}_b = 0.02334$, and $t = -0.63$. With df = 13, we see that $P > 2(0.25) = 0.50$ (software reports 0.541), so we have no evidence to suggest a straight-line relationship. While the relationship between these two variables is very strong, it is definitely not linear. See also the solution to Exercise 3.11.

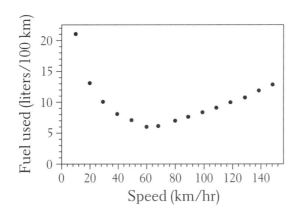

14.10 (a) r^2 is very close to 1, which means that nearly all the variation in steps per second is accounted for by foot speed. Also, the P-value for β is small.

(b) β (the slope) is this rate; the estimate is listed as the coefficient of "Speed," 0.080284. Using a $t(5)$ distribution: $0.080284 \pm (4.032)(0.0016) = 0.07383$ to 0.08674.

14.11 (a) The plot (below) shows a strong positive linear relationship. (b) β (the slope) is this rate; the estimate is listed as the coefficient of "year": 9.31868. (c) df $= 11$; $t^* = 2.201$; $9.31868 \pm (2.201)(0.3099) = 8.6366$ to 10.0008.

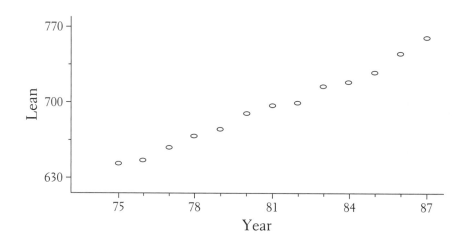

14.12 (a) One residual (51.32) may be a high outlier, but the stemplot does not show any other deviations from normality.

$$
\begin{array}{r|l}
-3 & 1 \\
-2 & 433 \\
-1 & 975532 \\
-0 & 999976663220 \\
0 & 0339 \\
1 & 01114899 \\
2 & 14 \\
3 & 3 \\
4 & \\
5 & 1
\end{array}
$$

(b) The scatter of the data points about the regression line (see Figure 14.1) varies to a certain extent as we move along the line, but the variation is not serious, as a residual plot shows. The other conditions can be assumed to be satisfied. (c) A prediction interval would be wider. For a fixed confidence level, the margin of error is always larger when we are predicting a single observation (a variable quantity) than when we are estimating the mean response. (d) We are 95% confident that when x (crying intensity) $= 25$, the corresponding value of y (IQ) will be between 91.85 and 165.33.

14.13 (a) The major difficulty is that the observations are not independent. The number of powerboat registrations for any year is related to the number of registrations for the previous year. The other conditions can be assumed to be satisfied.

(b) The confidence interval is (41.43, 49.59). The prediction interval is (33.35, 57.66). The confidence interval is more precise (i.e., narrower) since it is based on the mean of the observations, and the prediction interval is calculated for a single observation.

14.14 The number of points is so small that it is hard to judge much from the stemplot. The scatterplot of residuals vs. year does not suggest any problems. The regression in Exercise 14.11 should be fairly reliable.

```
-0 | 6
-0 | 55
-0 | 32
-0 |
 0 | 011
 0 | 22
 0 | 44
 0 | 7
```

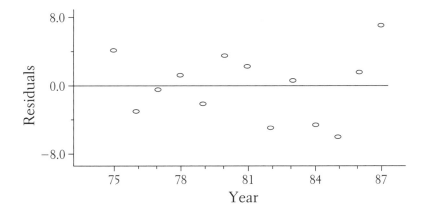

14.15 The scatterplot (below) shows a positive association. The regression line is $\hat{y} = 113.2 + 26.88x$; the linear relationship with body mass accounts for $r^2 = 74.8\%$ of the variation in metabolic rate. See also the solution to Exercise 3.12.

Minitab output (on the next page) reports $b = 26.879$ and $\mathrm{SE}_b = 3.786$; with df = 17, the critical value is $t^* = 1.740$, so the 90% confidence interval for β is $26.879 \pm (1.740)(3.786) = 20.29$ to 33.47 cal/kg. For each additional kilogram of mass, metabolic rate increases by about 20 to 33 calories.

The residuals are listed on the next page (in order, down the columns). A stemplot (on the next page) suggests that the distribution of residuals is right-skewed, and the largest residual may be an outlier. A scatterplot (on the next page) of the residuals against the explanatory variable gives some hint that the variation about the line is not constant (in violation of the regression assumptions). However, the three highest residuals account for most of that impression (as well as the skewness of the distribution), so these three individuals may need to be examined further.

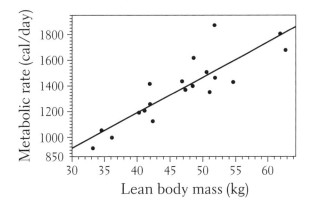

Minitab output

```
The regression equation is
Rate = 113 + 26.9 Mass

Predictor      Coef      Stdev     t-ratio          p
Constant      113.2      179.6        0.63      0.537
Mass         26.879      3.786        7.10      0.000

s = 133.1    R-sq = 74.8%    R-sq (adj) = 73.3%
```

Residuals

12.36	−7.37	−1	5
−137.83	−89.85	−1	332
−88.48	−48.16	−0	88
−155.74	−128.82	−0	42210
−20.78	11.52	0	1112
175.93	−139.66	0	6
−25.21	−16.56	1	
28.78	358.84	1	79
13.93	65.23	2	
191.85		2	
		3	
		3	5

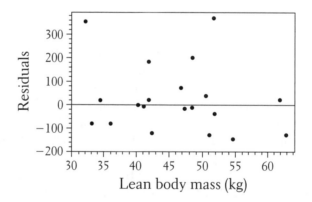

14.16 From the computer regression output for the years 1977–1994 (see below), we note that the regression equation is $\hat{y} = -35.2 + 0.113x$ and $s = 5.399$. The critical value for a 90% confidence interval with df = 17 is $t^* = 1.740$. The 90% confidence interval for mean response (mean number of manatees killed) at $x = 700$ is (40.60, 46.81).

```
The regression equation is
MANDTH = − 35.2 + 0.113 REG

Predictor         Coef       StDev            T          P
Constant       −35.179       7.696        −4.57      0.000
REG            0.11269      0.01262         8.93      0.000

S = 5.399   R-Sq = 83.3%        R-Sq (adj) = 82.3%
--------------------------------------------------------------
Fit       StDev Fit       90.0% CI              90.0% PI
43.70        1.78      (40.60, 46.81)       (33.78, 53.63)
```

14.17 (a) Stemplots and boxplots of the data show that both armspan and height are approximately normally distributed, with height slightly skewed right. It is reasonable to assume that the data are independent observations from normal populations; that, for given armspan, heights would be approximately normally distributed; and that the standard deviation σ of heights is the same for all values of armspan.

(b) 95% of the time, the prediction interval corresponding to armspan = 75 inches will capture the true height.

(c) The 95% confidence interval corresponding to armspan = 75 inches predicts the mean height for all those individuals with armspan = 75 inches. The prediction interval establishes a range for the prediction of one student with armspan = 75, while a confidence interval establishes a range of heights for the mean height of all students with armspan = 75. Since averages have less variation that individual observations, the confidence interval will be shorter.

14.18 Plot below. The regression equation is $\hat{y} = 560.65 - 3.0771x$; this and the plot (below) show that generally, the longer a child remains at the table, the fewer calories he or she will consume.

Software (output below) reports that $SE_b = 0.8498$; to compute this by hand, first note that the estimated standard deviation is $s = 23.40$ calories and that $\sqrt{\Sigma(x - \bar{x})^2}$ can be found by multiplying the standard deviation of x (Time) by $\sqrt{19}$. For df = 18, $t^* = 2.101$, so the 95% confidence interval is $b \pm t^*SE_b = -3.0771 \pm (2.101)(0.8498) = -4.8625$ to -1.2917 calories per minute.

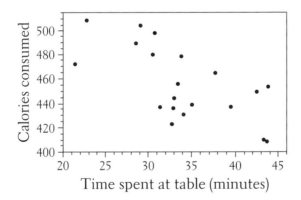

Time spent at table (minutes)

Minitab output

```
The regression equation is
Calories = 561 - 3.08 Time
Predictor        Coef      Stdev     t-ratio        P
Constant       560.65      29.37       19.09    0.000
Time          -3.0771     0.8498       -3.62    0.002

s = 23.40      R-sq = 42.1%       R-sq(adj) = 38.9%
```

14.19 (a) Stumps (the explanatory variable) should be on the horizontal axis; the plot shows a positive linear association. See also the solution to Exercise 3.45.

(b) The regression line is $\hat{y} = -1.286 + 11.894x$. Regression on stump counts explains 83.9% of the variation in the number of beetle larvae.

(c) Our hypotheses are $H_0: \beta = 0$ versus $H_a: \beta \neq 0$, and the test statistic is $t = 10.47$ (df = 21). The output shows $P = 0.000$, so we know that $P < 0.0005$ (as we can confirm from Table C); we have strong evidence that beaver stump counts help explain beetle larvae counts.

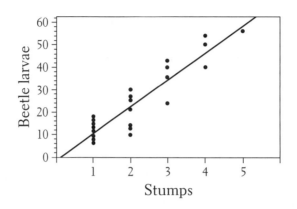

14.20 (a) The mean is $\bar{x} \doteq 0.00174$, and the standard deviation is $s \doteq 1.0137$. For a standardized set of values, the mean and standard deviation should be (up to rounding error) 0 and 1, respectively.*

(b) The stemplot (below) doesn't look particularly symmetric, but is not strikingly nonnormal (for such a small sample). In a set of 23 observations from a standard normal distribution, we expect most (95%) to be between -2 and 2, so -1.99 is quite reasonable.

(c) The plot of residuals versus stump counts (below) gives no cause for concern.

```
-1 | 965
-1 | 30
-0 | 7
-0 | 4422
 0 | 0224
 0 | 56789
 1 | 2233
```

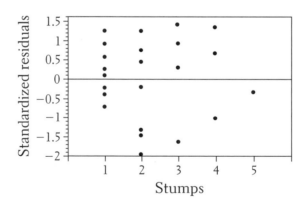

*Note to instructors: Most students do not need to know—but you should—that the process of finding standardized residuals is more complicated than simply finding the standard deviation of the set of residuals and then dividing each residual by that standard deviation. (The mean of the residuals will necessarily be 0.) Standardized residuals will generally have mean and standard deviation *close to*, but not exactly equal to, 0 and 1 (respectively). For example, with this data set, even working with unrounded standardized residuals, the mean is 0.0015107 (not 0) and the standard deviation is 1.0140 (not 1). A simpler example is the data set $(1, 1), (2, 1), (3, 2)$, for which the standardized residuals are $1, -1, 1$, with mean $1/3$ and standard deviation $\sqrt{4/3} \doteq 1.1547$.

14.21 (a) Below. U.S. returns (the explanatory variable) should be on the horizontal axis. Since both variables are measured in the same units, the same scale is used on both axes. See also the solution to Exercise 3.56.

(b) $t = b/\mathrm{SE}_b = 0.6181/0.2369 \doteq 2.609$. df = 25; since $2.485 < t < 2.787$, we know that $0.01 < P < 0.02$ (multiply the upper-tail probabilities by 2 since the alternative hypothesis is two-sided). Thus, we have fairly strong evidence for a linear relationship—that is, that the slope is nonzero.

(c) When $x = 15\%$, $\hat{y} = 5.683\% + 0.6181x = 14.95\%$. For estimating an individual y-value, we use the prediction interval: -19.65% to 49.56%.

(d) The width of the prediction interval is one indication that this prediction is not practically useful; another indication is that knowing the U.S. return accounts for only about $r^2 = 21.4\%$ of the variation in overseas returns.

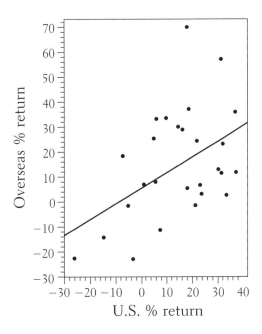

14.22 (a) The plot of residuals (below, left) suggests that variability about the line may not be constant for all values of x; it seems to increase from left to right.

(b) The stemplot (below, right) suggests that the distribution of residuals is right-skewed. The outlier is from 1986, when the overseas return was much higher than our regression predicts.

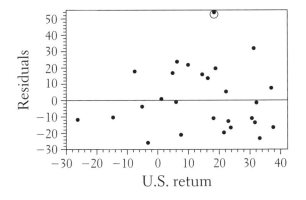

Residuals

-2	7420
-1	77432221
-0	4320
0	46
1	24568
2	02
3	0
4	
5	2

14.23 (a) Scatterplot below. Regression gives $\hat{y} = 166.5 - 1.099x$; the linear relationship explains about $r^2 = 20.9\%$ of the variation in yield.

(b) The t statistic for testing $H_0: \beta = 0$ vs. $H_a = \beta < 0$ is $t = -1.92$; with df $= 14$, the P-value is 0.0375 (Table C tells us that $0.025 < P < 0.05$). We have some evidence that weeds influence corn yields, but it is not strong enough to meet the usual standards of statistical significance.

(c) The small value of r^2 and the lack of significance of the t test indicate that this regression has little predictive use. When $x = 6$, $\hat{y} = 159.9$ bu/acre; the 95% confidence interval (given by Minitab, below) is 154.4 to 165.3 bu/acre. [Up to rounding error, this agrees with the "hand-computed" value, with $t^* = 2.145$ and $\text{SE}_{\hat{\mu}} = 2.54$: $159.9 \pm (2.145)(2.54)$.] The width of this interval is another indication that the model has little practical use.

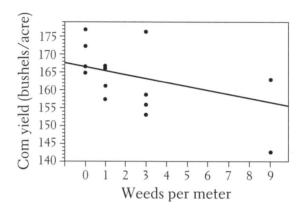

Minitab output

```
The regression equation is
Corn = 166 - 1.10 Weeds

Predictor        Coef       Stdev      t-ratio           P
Constant      166.483       2.725        61.11       0.000
Weeds         -1.0987      0.5712        -1.92       0.075

s = 7.977      R-sq = 20.9%        R-sq (adj) = 15.3%

----------------------------(output continues)----------------------------

       Fit     Stdev. Fit        95.0% C.I.            95.0% P.I.
    159.89            2.54    (154.44, 165.34)     (141.93, 177.85)
```

14.24 df $= 21$, so $t^* = 1.721$; the 90% confidence interval is $b \pm t^* \text{SE}_b = -9.6949 \pm (1.721)(1.8887) = -12.9454$ to -6.4444 bpm per minute. With 90% confidence, we can say that for each 1-minute increase in swimming time, pulse rate drops by 6 to 13 bpm.

14.25 (a) The regression line is $\hat{y} = 479.9 - 9.6949x$; with $x = 34.3$ minutes, this agrees with the output (up to rounding). The prediction interval is appropriate for estimating one value (as opposed to the mean of many values): 135.79 to 159.01 bpm.

(b) Using df $= 21$, we find $t^* = 1.721$; this would give the interval $147.40 \pm (1.721)(1.97) = 144.01$ to 150.79, which agrees with the computer output (up to rounding error).

14.26 (a) Perch #143 lies slightly above the overall linear pattern, but does not appear to be too far out of place on the width versus length plot (see next page). Since both variables are measured in centimeters, the same scale is used on both axes.

(b) Regression (see Minitab output below) gives $\hat{y} = -0.6522 + 0.1823x$ cm.

(c) When $x = 27$, $\hat{y} \doteq 4.27$ cm. We have df = 54, so we use 50 degrees of freedom in Table C, which gives $t^* = 2.009$. The 95% confidence interval for μ_y is $4.27 \pm (2.009)(0.0552) = 4.16$ to 4.38 cm.

(d) A stemplot of the residuals (below, left) reveals a high outlier (which came from perch #143) and a less extreme low value (from fish #149, which is unusually slender for its width). A plot of residuals versus length (below, right) suggests that there may be more variability in width for larger lengths (although much of this impression may be due to the two extreme residuals and the fact that we have fewer observations for small fish). These two issues might give us reason to be hesitant about using inference procedures.

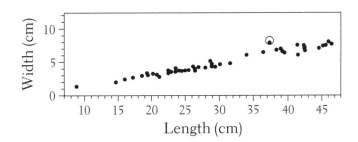

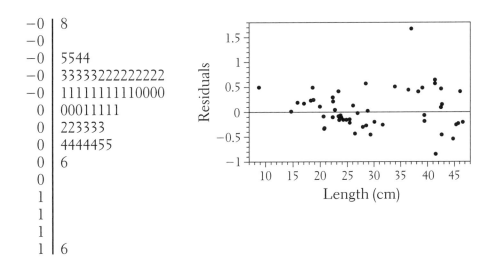

Minitab output

```
The regression equation is
width = - 0.652 + 0.182 length
Predictor        Coef        Stdev      t-ratio          p
Constant      -0.6522      0.1751        -3.72      0.000
length       0.182322     0.005642        32.32      0.000

s = 0.3987    R-sq = 95.1%    R-sq(adj) = 95.0%

------------------------(output continues)------------------------

    Fit     Stdev.Fit      95.0% C.I.           95.0% P.I.
 4.2705       0.0552    (4.1597, 4.3812)    (3.4633, 5.0777)
```

14.27 (a) The plot (below, left) shows a fairly strong curved pattern (weight increases with length). Two points are circled; these fish—#143 and #149, the two fish noted in the previous problem—stray the most from the curve, but probably would not be considered outliers.

(b) We would expect weight to increase roughly linearly with volume (if we double volume, we double weight; if we triple the volume, we triple the weight, etc.). When all dimensions (length, width, and height) are doubled, the *volume* of an object increases by a factor of $8 = 2^3$. Similarly, if we triple all dimensions, volume (and, approximately, weight) increases by a factor of $27 = 3^3$. It then makes sense that the cube root (i.e., the one-third power) of the weight increases at an approximately linear rate with length.

(c) The second plot (below, right) also shows a strong positive association, but this is much more linear. There are no particular outliers.

(d) The correlations reflect the increased linearity of the second plot: Using the original weight variable, $r^2 = 0.9207$; with weight$^{1/3}$, $r^2 = 0.9851$.

(e) Regression gives $\hat{y} = -0.3283 + 0.2330x$; $\hat{y} = 5.9623$ when $x = 27$ cm. We have df $= 54$, so we use 50 degrees of freedom in Table C, which gives $t^* = 2.009$. The 95% confidence interval for μ_y is $5.9623 \pm (2.009)(0.0382) = 5.886$ to 6.039 g$^{1/3}$.

(f) The stemplot shows no gross violations of the assumptions except for the high outlier for fish #143. As we saw in the previous problem, the scatterplot suggests that variability in weight may be greater for larger lengths. Neither of these violations is too severe, but together they might be a cause for concern. However, dropping fish #143 (which changes the regression line only slightly, to $\hat{y} = -0.2975 + 0.2313x$) seems to alleviate both these problems (to some degree, at least), so regression should be safe.

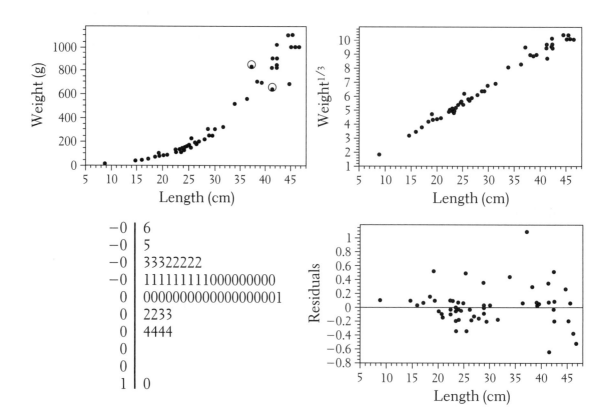

Minitab output

```
The regression equation is
cuberoot = − 0.328 + 0.233 length

Predictor        Coef        Stdev     t-ratio         P
Constant      −0.3283       0.1211       −2.71     0.009
length       0.232986     0.003902       59.72     0.000


s = 0.2757    R-sq = 98.5%    R-sq(adj) = 98.5%


------------------------(output continues)-----------------------


    Fit    Stdev.Fit      95.0% C.I.            95.0% P.I.
 5.9623       0.0382    (5.8857, 6.0389)    (5.4041, 6.5205)
```

15

Analysis of Variance

15.1 (a) $F^* = 3.68$. (b) Not significant at either 10% or 5% (in fact, $P = 0.1166$).

15.2 (a) Significant at 5%, but not at 1%. (b) Between 0.025 and 0.05 ($P = 0.0482$).

15.3 $F = 10.57$; P-value (for the two-sided alternative) is between 0.02 and 0.05 ($P = 0.0217$) — so this is significant at 5% but not at 1%.

15.4 Test: H_0: $\sigma_{\text{skilled}} = \sigma_{\text{novice}}$ vs. H_a: $\sigma_s \neq \sigma_{n.}$ $F = 4.007$; P-value is between 0.05 and 0.10 ($P = 0.0574$). This does not represent very strong evidence in favor of H_a.

15.5 To test H_0: $\sigma_1 = \sigma_2$ vs. H_a: $\sigma_1 \neq \sigma_2$, we find $F = (17.71)^2/(6.01)^2 \approx 8.68$. From an $F(9, 9)$ distribution, the P-value for the two-sided alternative is between 0.002 and 0.02 (software reports $P = 0.0036$). This is significant evidence of unequal standard deviations.

15.6 To test H_0: $\sigma_1 = \sigma_2$ vs. H_a: $\sigma_1 \neq \sigma_2$, we find $F = (5.44)^2/(5.05)^2 \approx 1.16$. From an $F(120, 100)$ distribution, the P-value for the two-sided alternative is greater than 0.20 (software reports $P = 0.375$ from an $F(161, 132)$ distribution). This gives us little reason to believe that the standard deviations are unequal.

15.7 $F = 1.5443$; the P-value is 0.3725, so the difference is not statistically significant.

15.8 (a) The $F(2, 25)$ curve begins at $(0, 1)$ and decreases, approaching the x-axis asymptotically.

(b) The $F(5, 50)$ curve begins near the origin, rises to a peak and then decreases, approaching the x-axis asymptotically.

(c) $F(12, 100)$ reaches a higher peak, which is slightly to the right of the peak for $F(5, 50)$.

(e) As the sample sizes increase, the F-distribution curve becomes more symmetric and looks more like a normal distribution.

15.9 (a) The stemplots (see next page) show no extreme outliers or skewness.

(b) The means suggest that a dog reduces heart rate, but being with a friend appears to raise it.

(c) $F = 14.08$ and $P = 0.000$ (meaning $P < 0.0005$), which means we reject H_0: $\mu_P = \mu_F = \mu_C$ in favor of H_a: at least one mean is different. Based on the confidence intervals, it appears that the mean heart rate is lowest when a pet is present (although this interval overlaps the control interval) and is highest when a friend is present (although again, this interval overlaps the control interval).

```
   Dog          Friend       Alone
 5 | 9        5 |          5 |
 6 | 4        6 |          6 | 3
 6 | 5999     6 |          6 |
 7 | 0002     7 |          7 | 13
 7 | 6        7 | 7        7 | 58
 8 | 0        8 | 023      8 | 0
 8 | 56       8 | 78       8 | 555778
 9 |          9 | 012      9 | 02
 9 | 8        9 | 78       9 | 9
10 |         10 | 0112    10 |
```

15.10 (a) Stemplots (below) suggest that yields first increase with plant density, then decrease. (b) H_0: $\mu_1 = \mu_2 = \mu_3 = \mu_4 = \mu_5$ (all plant densities give the same mean yield per acre) versus H_a: not all means are the same. (c) Mean yields: 131.03, 143.15, 146.22, 143.07, and 134.75 bushels per acre. With $F = 0.50$ and $P = 0.736$, we conclude that the differences are not significant. (d) The sample sizes were small, which means there is a lot of potential variation in the outcome. (That is why the confidence intervals shown in the printout are very wide.)

```
  12,000      16,000      20,000      24,000      28,000
11 | 38     11 |        11 |        11 |        11 | 9
12 |        12 | 1      12 |        12 |        12 |
13 |        13 | 5      13 | 0      13 | 58      13 |
14 | 3      14 |        14 | 0      14 |        14 |
15 | 0      15 | 0      15 | 0      15 | 6      15 | 1
16 |        16 | 7      16 | 5      16 |        16 |
```

15.11 (a) I, the number of populations, is 3; the sample sizes from each population are $n_1 = n_2 = n_3 = 15$; the total sample size is $N = 45$.
(b) Numerator ("Group"): $I - 1 = 2$, denominator ("Error"): $N - I = 42$.
(c) Since $F > 9.22$, the largest critical value for an $F(2, 25)$ distribution in Table D, we conclude that $P < 0.001$.

15.12 (a) I, the number of populations, is 5; the sample sizes are $n_1 = 4$, $n_2 = 4$, $n_3 = 4$, $n_4 = 3$, and $n_5 = 2$; the total sample size is $N = 17$.
(b) Numerator ("Rate"): $I - 1 = 4$, denominator ("Error"): $N - I = 12$. (c) Since $F < 2.48$, the smallest critical value for an $F(4, 12)$ distribution in Table D, we conclude that $P > 0.100$.

15.13 (a) Populations: tomato varieties; response variable: yield. $I = 4$, $n_1 = n_2 = n_3 = n_4 = 10$, and $N = 40$, so there are $I - 1 = 3$ and $N - I = 36$ degrees of freedom.
(b) Populations: consumers (responding to different package designs); response variable: attractiveness rating. $I = 6$, $n_1 = n_2 = n_3 = n_4 = n_5 = n_6 = 120$, and $N = 720$, so there are $I - 1 = 5$ and $N - I = 714$ degrees of freedom.
(c) Populations: dieters (under different diet programs); response variable: weight change after 6 months. $I = 3$, $n_1 = n_2 = 10$, $n_3 = 12$, and $N = 32$, so there are $I - 1 = 2$ and $N - I = 29$ degrees of freedom.

15.14 (a) Yes: $\frac{\text{largest } s}{\text{smallest } s} = \frac{9.970}{8.341} \doteq 1.20$. (b) Yes: $\frac{\text{largest } s}{\text{smallest } s} = \frac{22.27}{11.44} \doteq 1.95$.

15.15 (a) The biggest difference is that single men earn considerably less than men who have been or are married. Widowed and married men earn the most; divorced men earn about \$1,300 less (on the average), and single men are \$4,000 below that.

(b) Yes: $\frac{8119}{5731} \doteq 1.42$.

(c) 3 and 8231.

(d) The sample sizes are so large that even small differences would be found to be significant; we have some fairly large differences.

(e) No—single men are likely to be younger than men in the other categories. This means that typically they have less experience, and have been with their companies less time than the others, and so have not received as many raises, etc.

15.16 (a) $H_0: \mu_{r1} = \mu_{r2} = \mu_{r3}$ (all class rank means are same) vs. H_a: not all means are the same.

(b) 2 and 253 (for all three tests).

(c) Yes: $\frac{10.8}{10.5} = 1.03$, $\frac{1.31}{1.17} = 1.12$, and $\frac{.55}{.40} = 1.375$. Comparing to $F(2,200)$ critical values, we find $P_{rank} > 0.100$, P_{sem} is between 0.010 and 0.025, and $P_{grade} < 0.001$.

(d) Mean high school class rank varies little between the groups. Regarding the other two variables, there appears to be little difference between the CS and Sci/Eng majors. However, "semesters of HS math" and "average grade in HS math" both show a significant difference between CS/Sci/Eng majors and those in the "Other" category. On the average, the first two groups had about one half-semester more math, and had grades about 0.25 higher.

15.17 This exercise is based on Activity 15. Answers will vary.

15.18 (a) Multiply the ith standard error by $\sqrt{n_i}$ to find the standard deviation: $s_{cold} = (8.08)(\sqrt{16})$ $= 32.32$, $s_{neutral} = (5.61)(\sqrt{38}) = 34.58$, $s_{hot} = (4.10)(\sqrt{75}) = 35.51$. This satisfies the rule of thumb for ANOVA: $s_{max}/s_{min} = 35.51/32.32 \approx 1.10$.

(b) We have $I = 3$, $N = 16 + 38 + 75 = 129$. Thus,

$$\bar{x} = \frac{(16)(28.89) + (38)(32.93) + (75)(32.27)}{129} \approx 32.045$$

$$MSG = \frac{(16)(28.89 - 32.045)^2 + (38)(32.93 - 32.045)^2 + (75)(32.27 - 32.045)^2}{3 - 1} \approx 96.412$$

$$MSE = \frac{(15)(32.32)^2 + (37)(34.58)^2 + (74)(35.51)^2}{129 - 3} \approx 1216.06$$

$$F = \frac{96.412}{1216.06} \approx 0.08$$

With df $= 2$ and 126, $P > 0.100$, so there is no reason to believe that nest temperature affects the python's weight at hatching.

15.19 (a) $s_{cold} = (5.67)\sqrt{16} = 22.68$, $s_{neutral} = (4.24)\sqrt{38} = 26.14$, and $s_{hot} = (2.70)\sqrt{75} = 23.38$. This easily satisfies our rule of thumb: $\frac{26.14}{22.68} \doteq 1.15$.

(b) We have $I = 3$ and $N = 129$, so

$$\bar{x} = \frac{(16)(6.40) + (38)(5.82) + (75)(4.30)}{129} \doteq 5.008$$

$$MSG = \frac{(16)(6.40 - 5.008)^2 + (38)(5.82 - 5.008)^2 + (75)(4.30 - 5.008)^2}{3 - 1} \doteq 46.83$$

$$MSE = \frac{(15)(22.68)^2 + (37)(26.14)^2 + (74)(23.38)^2}{129 - 3} \doteq 582.9$$

$$F = \frac{46.83}{582.9} \doteq 0.08$$

With df = 2 and 126, $P > 0.100$, so we have no reason to doubt the null hypothesis; that is, there is no evidence that nest temperature affects propensity to strike.

15.20 (a) We have $I = 5$ and $N = 17$, so

$$MSE = \frac{(3)(18.09)^2 + (3)(19.79)^2 + (3)(15.07)^2 + (2)(11.44)^2 + (1)(22.27)^2}{17 - 5} \doteq 299.64$$

$$\bar{x} = \frac{(4)(131.03) + (4)(143.15) + (4)(146.22) + (3)(143.07) + (2)(134.75)}{17} \doteq 140.02$$

$$MSG = \frac{(4)(131.03 - 140.02)^2 + \cdots + (2)(134.75 - 140.02)^2}{5 - 1} = \frac{599.68}{4} = 149.9$$

MSE and MSG agree with Minitab's output (except for rounding).
(b) Use $t^* = 1.782$ from a $t(12)$ distribution: $146.22 \pm (1.782)(17.31/\sqrt{4}) = 130.80$ to 161.64 bushels per acre.

15.21 (a) The distribution for perch has an extreme high outlier (20.9); the bream distribution has a mild low outlier (12.0). Otherwise there is no strong skewness. (b) Bream: 12.0, 13.6, 14.1, 14.9, 15.5; Perch: 13.2, 15.0, 15.55, 16.675, 20.9; Roach: 13.3, 13.925, 14.65, 15.275, 16.1. The most important difference seems to be that perch are larger than the other two fish. It also appears that (typically) bream *may* be slightly smaller than roach.

Bream		Perch		Roach	
12	0	12		12	
12		12		12	
13	333344	13	2	13	3
13	5677788889	13	69	13	6799
14	111233	14	3	14	0013
14	78899	14	55667889	14	677
15	001113	15	000000011112344	15	122344
15	5	15	56788999	15	6
16		16	0112333	16	1
16		16	8	16	
17		17	003	17	
17		17	56667789	17	
18		18	1	18	
18		18		18	
19		19		19	
19		19		19	
20		20		20	
20		20	9	20	

15.22 (a) H_0: $\mu_1 = \mu_2 = \mu_3$—the mean weights of the three types are equal. (b) $F = 29.92$ and $P < 0.001$. (c) Perch seem to actually be different in weight from the other two species; the mean weight of bream and roach may not differ greatly (the two confidence intervals overlap).

15.23 (a) I, the number of populations, is 3; the sample sizes from the 3 populations are $n_1 = 35$, $n_2 = 55$ (after discarding the outlier), and $n_3 = 20$; the total sample size is $N = 110$.

(b) numerator ("factor"): $I - 1 = 2$, denominator ("error"): $N - I = 107$.

(c) Since $F > 7.41$, the largest critical value for an $F(2,100)$ distribution in Table D, we conclude that $P < 0.001$.

15.24 Yes: $\frac{\text{largest } s}{\text{smallest } s} = \frac{1.186}{0.770} = 1.54$.

15.25 (a) MSE $= \frac{1}{107}[(34)(0.770)^2 + (54)(1.186)^2 + (19)(0.780)^2] = \frac{107.67}{107} = 1.0063$; $s_p = \sqrt{\text{MSE}} = 1.003$.

(b) Use $t^* = 1.984$ from a $t(100)$ distribution (since $t(107)$ is not available): $15.747 \pm t^* s_p / \sqrt{55} = 15.479$ to 16.015. Using software, we find that for a $t(107)$ distribution, $t^* = 1.982$; this rounds to the same interval.

(c) $\bar{x} = \frac{1}{110}[(35)(14.131) + (55)(15.747) + (20)(14.605)] = 15.025$. MSG $= \frac{1}{2}[(35)(14.131 - 15.025)^2 + (55)(15.747 - 15.025)^2 + (20)(14.605 - 15.025)^2] = 30.086$.

(d) $F = \frac{30.086}{1.003} = 29.996$—reasonably close to Minitab's output.

P-value $=$ Fcdf $(29.996, 1E99, 2,107) = 4.55 \times 10^{-11}$, or essentially 0.

15.26 We wish to test H_0: $\sigma_1 = \sigma_2$ vs. H_a: $\sigma_1 \neq \sigma_2$, where σ_1, σ_2 are the standard deviations of weights in kilograms for skilled and novice rowers, respectively. From the output in Exercise 11.48, we see that $s_1 = 6.10$ and $s_2 = 9.04$. No information is given on the distribution of weights within each sample (in fact, we do not know the original data at all), but the assumption of normality is reasonable because we are dealing with a physical characteristic of a large population. TI-83 output for the two-sample F test is given below. (Note that we let the first standard deviation be the larger of the two s's above.) Test statistic $F = \frac{(9.04)^2}{(6.1)^2} = 2.196$. Using the $F(7, 9)$ distribution in Table D, we see that the P-value is greater than .20 (using software, the exact value is 0.2697). The difference between σ_1, σ_2 is not significant.

```
2-SampFTest          2-SampFTest
Inpt:Data Stats      σ1≠σ2
Sx1:9.04             F=2.196226821
n1:8                 P=.2695882678
Sx2:6.1              Sx1=9.04
n2:10                Sx2=6.1
σ1:█  <σ2  >σ2       ↓n1=8
Calculate Draw
```

15.27 (a) We wish to test H_0: $\sigma_1 = \sigma_2$ vs. H_a: $\sigma_1 \neq \sigma_2$, where σ_1, σ_2 are the standard deviations of Chapin test scores for female and male liberal-arts majors, respectively. The value of the test statistic $F = \frac{(5.44)^2}{(5.05)^2} = 1.16$. Using the $F(200, 120)$ distribution in Table D, we see that the P-value is greater than .20 (software yields $P = 0.3754$). There is no evidence that the standard deviations are different.

(b) No. F procedures are not robust against nonnormality, even with large samples.

15.28 (a) Populations: nonsmokers, moderate smokers, and heavy smokers; response variable: hours of sleep per night. $I = 3$, $n_1 = n_2 = n_3 = 200$, and $N = 600$; 2 and 597 df.

(b) Populations: different concrete mixtures; response variable: strength. $I = 5$, $n_1 = \cdots = n_5 = 6$, and $N = 30$; 4 and 25 df.

(c) Populations: teaching methods; response variable: test scores. $I = 4$, $n_1 = n_2 = n_3 = 10$, $n_4 = 12$, and $N = 42$; 3 and 38 df.

15.29 (a) Only Design A would allow use of one-way ANOVA because it produces four independent sets of numbers. The data resulting from Design B would be dependent (a subject's responses to the first list would be related to that same subject's responses to the other lists), so that ANOVA would not be appropriate for comparison.

(b) The ANOVA test statistic is $F = 4.92$ (df 3 and 92), which has $P = 0.003$, so there is strong evidence that the means are not all the same. In particular, list 1 seems to be the easiest, and lists 3 and 4 are the most difficult.

15.30 (a) The stemplots (below) appear to be acceptable; they show no extreme outliers or strong skewness (given the small sample sizes).

(b) The standard deviations are quite similar and easily satisfy our rule of thumb: $\frac{5.761}{4.981} \doteq 1.16$. The means appear to suggest that logging reduces the number of trees per plot and that recovery is slow (the 1-year-after and 8-years-after numbers are similar). $F = 11.43$ with df $= 2$ and 30 has $P < 0.001$, so we conclude that these differences are significant; the number of trees per plot really is lower in logged areas.

Never logged		1 year ago		8 years ago
0		0	2	0	4
0		0	9	0	
1		1	2244	1	22
1	699	1	57789	1	5889
2	0124	2	0	2	22
2	7789	2		2	
3	3	3		3	

15.31 (a) The table is given in the Minitab output below; since $\frac{4.500}{3.529} \doteq 1.28$, ANOVA should be safe. The means appear to suggest that logging reduces the number of species per plot and that recovery takes more than 8 years. (b) ANOVA gives $F = 6.02$ with df $= 2$ and 30, so $P < 0.010$ (software gives 0.006). We conclude that these differences are significant; the number of species per plot really is lower in logged areas.

Minitab output

```
Analysis of Variance on Species
Source     DF        SS        MS       F         P
Code        2     204.4     102.2    6.02     0.006
Error      30     509.2      17.0
Total      32     713.6

                              Individual 95% CIs For Mean
                              Based on Pooled StDev
Level      N      Mean     StDev    ---------+---------+---------+---------
    1     12    17.500     3.529                          (---------*---------)
    2     12    11.750     4.372    (---------*---------)
    3      9    13.667     4.500          (---------*---------)
                                    ---------+---------+---------+---------
Pooled StDev =  4.120                   12.0      15.0      18.0
```

15.32 (a) The data suggest that the presence of too many nematodes reduces growth. Table and stemplots below.

(b) $H_0: \mu_1 = \cdots = \mu_4$ (all mean heights are the same) vs. H_a: not all means are the same. This ANOVA tests whether nematodes affect mean plant growth.

(c) Minitab output is shown below. The first two levels (0 and 1000 nematodes) do not appear to be significantly different, nor do the last two. However, it does appear that somewhere between 1000 and 5000 nematodes, the tomato plants begin to feel the effects of the worms, and are hurt by their presence.

Nematodes	n_i	$\bar{x}_i$	s_i
0	4	10.650	2.053
1000	4	10.425	1.486
5000	4	5.600	1.244
10,000	4	5.450	1.771

```
      0              1000           5000           10,000

   3  |           3  |           3  |           3  | 2
   4  |           4  |           4  | 6         4  |
   5  |           5  |           5  | 04        5  | 38
   6  |           6  |           6  |           6  |
   7  |           7  |           7  | 4         7  | 5
   8  |           8  | 2         8  |           8  |
   9  | 12        9  |           9  |           9  |
  10  | 8        10  |          10  |          10  |
  11  |          11  | 113      11  |          11  |
  12  |          12  |          12  |          12  |
  13  | 5        13  |          13  |          13  |
```

Minitab output

```
Analysis of Variance
Source     DF          SS          MS          F           P
Factor      3      100.65       33.55       12.08       0.001
Error      12       33.33        2.78

Total      15      133.97
                                            Individual 95% CIs for Mean
                                            Based on Pooled StDev
Level      N        Mean      StDev    ---------+---------+---------+---------+
0          4      10.650      2.053                             (-------*-------)
1000       4      10.425      1.486                             (-------*-------)
5000       4       5.600      1.244       (--------*--------)
10000      4       5.450      1.771       (--------*-------)
                                        -------+---------+---------+---------+
POOLED STDEV = 1.667                       5.0       7.5      10.0      12.5
```

15.33 (a) $t = -0.34135$ with df $= 12$; $P = 0.7387$. (b) $\bar{x} = (2.344)$, MSG $= 0.02652$, MSE $= 0.25411$, and $F = 0.10436$; $P = 0.7491$. (c) The two P-values differ by about 0.01—an unimportant difference in most cases.

15.34 Using the means and standard deviations from problem 15.32(a): $\bar{x} = \frac{1}{16}[(4)(10.650) + (4)(10.425) + (4)(5.600) + (4)(5.450)] = 8.031$; all other values can be confirmed from the Minitab output in problem 15.32(c). Table D places the P-value at less than 0.001; software gives $P = 0.0006$.

15.35 (a) Stemplots of the basal, DRTA, and stratified data (left to right below) reveal no severe outliers.

```
 4 | 0            6 | 0           4 | 00
 5 |              7 | 000         5 | 0
 6 | 0            8 | 00000       6 | 000
 7 | 0            9 | 0000        7 | 000
 8 | 000         10 | 000         8 | 00
 9 | 000         11 |             9 | 0
10 | 0           12 | 00         10 |
11 | 0           13 | 00         11 | 000
12 | 0000000     14 |           12 | 00
13 | 0           15 | 0         13 | 000
14 | 0           16 | 0         14 | 00
15 | 0
16 | 0
```

We test H_0: the mean scores for all treatment groups are the same vs. H_a: the mean scores are not all the same, using ANOVA. The ANOVA output (see below) does not reveal any significant difference among the means ($P = 0.329$).

One-way Analysis of Variance

```
Analysis of Variance
Source      DF         SS        MS        F         P
Factor       2      20.58     10.29     1.13     0.329
Error       63     572.45      9.09
Total       65     593.03

                                   Individual 95% CIs for Mean
                                   Based on Pooled StDev
Level     N      Mean     StDev    -----+----------+----------+----------+-
BASAL    22    10.500     2.972                   (-----------*-----------)
DRTA     22     9.727     2.694          (-----------*-----------)
STRAT    22     9.136     3.342    (-----------*-----------)
                                   -----+----------+----------+----------+-
Pooled StDev = 3.014                   8.4        9.6       10.8       12.0
```

(b) Figure 15.14 suggests that the DRTA and Stratified groups had, on average, slightly higher COMP scores than the Basal group. (Note that the Basal and DRTA boxes and the Strat and DRTA boxes in Figure 15.14 just barely overlap). Also, in Figure 15.13, the means of the DRTA and Stratified groups appear to be somewhat larger than the Basal mean. The ANOVA table in Figure 15.15 reveals that there is a significant difference among the mean COMP scores for the three groups ($F = 4.48$, $P = 0.0152$). The contrast table in Figure 15.15 suggests a possible source of this significant difference: while the difference between DRTA and Stratified does not appear to be significant ($P = 0.2020$), the "Basal vs. DRTA and Stratified" contrast *is* significant ($P = 0.0088$). The output corroborates our subjective observation in Figures 15.13 and 15.14 that the Basal COMP scores are significantly lower than both the DRTA and the Stratified COMP scores.

16

Multiple Linear Regression

16.1 There are no clear, strong patterns. The GPA/SATM plot suggests a slight positive association, but it is weakened by the two low SATM scores which do not follow the pattern.

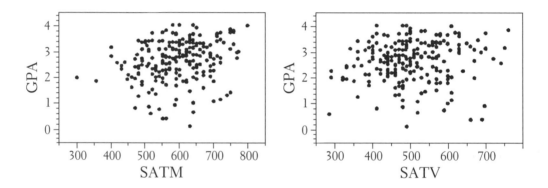

16.2 All of the plots display lots of scatter; high school grades seem to be poor predictors of college GPA. Of the three, the math plot seems to most strongly suggest a positive association, although the association appears to be quite weak, and almost nonexistent for HSM < 5. We also observe that scores below 5 are unusual for all three high school variables, and could be considered outliers and influential.

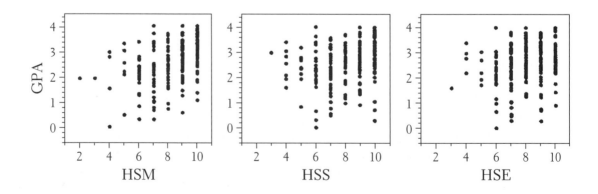

16.3 In the table (see facing page), two *IQR*s are given; those in parentheses are based on quartiles reported by Minitab, which computes quartiles in a slightly different way from this text's method.

None of the variables show striking deviations from normality in the quantile plots (not shown). Taste and H2S are slightly right-skewed, and Acetic has two peaks. There are no outliers.

	$\bar{x}$	M	s	IQR
Taste	24.53	20.95	16.26	23.9 (or 24.58)
Acetic	5.498	5.425	0.571	0.656 (or 0.713)
H2S	5.942	5.329	2.127	3.689 (or 3.766)
Lactic	1.442	1.450	0.3035	0.430 (or 0.4625)

Taste		Acetic		H2S		Lactic	
0	00	4	455	2	9	8	6
0	556	4	67	3	1268899	9	9
1	1234	4	8	4	17799	10	689
1	55688	5	1	5	024	11	56
2	011	5	2222333	6	11679	12	5599
2	556	5	444	7	4699	13	013
3	24	5	677	8	7	14	469
3	789	5	888	9	025	15	2378
4	0	6	0011	10	1	16	38
4	7	6	3			17	248
5	4	6	44			18	1
5	67					19	09
						20	1

16.4 The plots all show positive associations between the variables. The correlations and *P*-values (in parentheses) are on the next page; all are positive (as expected) and significantly different from 0. [Recall that the *P*-values are correct if the two variables are normally distributed, in which case $t = r\sqrt{n-2}/\sqrt{1-r^2}$ has a $t(n-2)$ distribution if $\rho = 0$.]

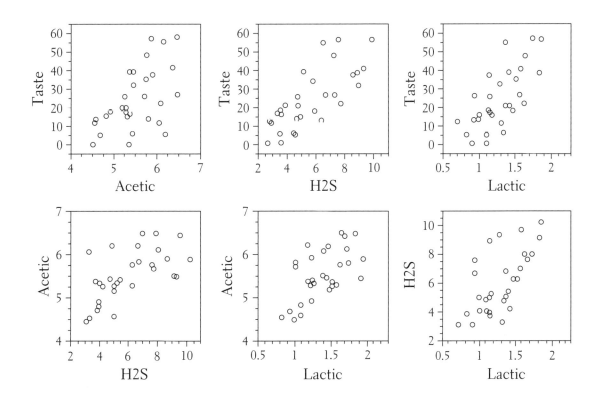

	Taste	Acetic	H2S
Acetic	0.5495 (0.0017)		
H2S	0.7558 (<0.0001)	0.6180 (0.0003)	
Lactic	0.7042 (<0.0001)	0.6038 (0.0004)	0.6448 (0.0001)

16.5 The regression equation is $\widehat{\text{Taste}} = -61.5 + 15.6$ Acetic; the coefficient of Acetic has $t = 3.48$, which is significantly different from 0 ($P = 0.002$). The regression explains $r^2 = 30.2\%$ of the variation in Taste.

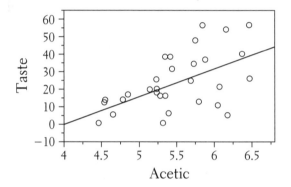

Based on stem- and quantile plots (not shown), the residuals seem to have a normal distribution. Scatterplots (below) reveal positive associations between residuals and both H2S and Lactic. Further analysis of the residuals shows a stronger positive association between residuals and Taste, while the plot of residuals vs. Acetic suggests greater scatter in the residuals for large Acetic values.

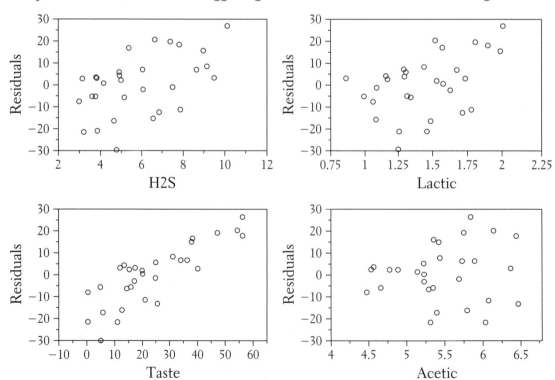

16.6 Regression gives $\widehat{\text{Taste}} = -9.79 + 5.78$ H2S. The coefficient of H2S has $t = 6.11$ ($P < 0.0005$); it is significantly different from 0. The regression explains $r^2 = 57.1\%$ of the variation in Taste.

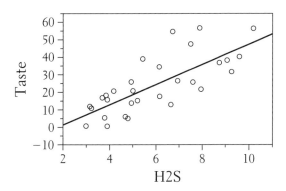

Based on stem- and quantile plots (not shown), the residuals may be slightly skewed, but do not differ greatly from a normal distribution. Scatterplots (below) reveal weak positive associations between residuals and both Acetic and Lactic. Further analysis of the residuals shows a moderate positive association between residuals and Taste, while the plot of residuals vs. H2S suggests greater scatter in the residuals for large H2S values.

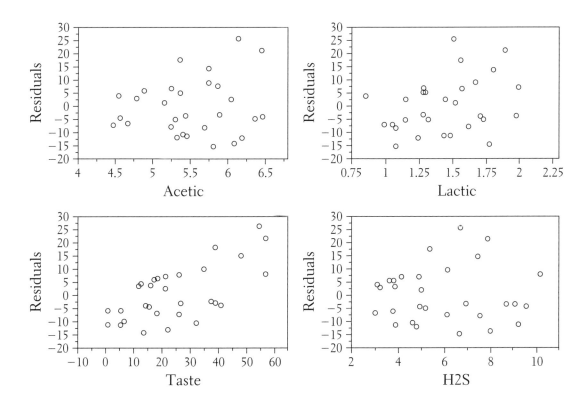

16.7 Regression gives $\widehat{\text{Taste}} = -29.9 + 37.7$ Lactic. The coefficient of Lactic has $t = 5.25$ ($P < 0.0005$); it is significantly different from 0. The regression explains $r^2 = 49.6\%$ of the variation in Taste.

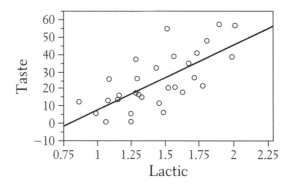

Based on stem- and quantile plots (not shown), the residuals seem to have a normal distribution. Scatterplots reveal a moderately strong positive association between residuals and Taste, but no striking patterns for residuals vs. the other variables.

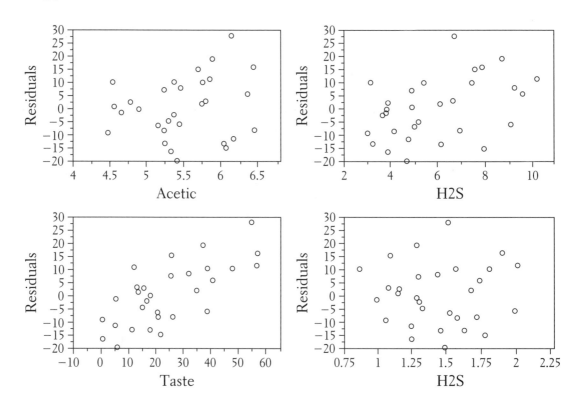

16.8 (a) The plot (see next page) reveals no outliers or unusual points. (b) The regression equation is $\hat{y} = -2.80 + 0.0387x$. (c) $t = 16.10$ (df = 17); since $P < 0.0005$, we reject H_0 and conclude that linear regression on HR is useful for predicting VO2. (d) When $x = 95$, we have $\hat{y} = 0.8676$ and $SE_{\hat{y}} = 0.1205\sqrt{1 + \frac{1}{19} + \frac{(95 - 107)^2}{2518}} = 0.1269$, so the 95% prediction interval is $0.8676 \pm (2.110)(0.1269)$, or 0.5998 to 1.1354. When $x = 110$, we have $\hat{y} = 1.4474$ and $SE_{\hat{y}} = 0.1205\sqrt{1 + \frac{1}{19} + \frac{(110 - 107)^2}{2518}} = 0.1238$, so the interval is $1.4474 \pm (2.110)(0.1238) = 1.1861$ to 1.7086. A portion of the Minitab output that shows these intervals is reproduced below. (e) It depends on how accurately they need to know VO2; the regression equation predicts only the subject's *mean* VO2 for a given heart rate, and the intervals in (d) reveal that a particular *observation* may vary quite a bit from that mean.

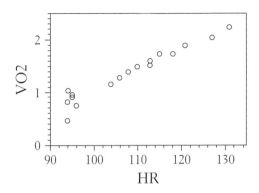

Output from Minitab

```
Fit          Stdev.Fit          95.0% C.I.                  95.0% P.I.
0.8676         0.0399      ( 0.7834,   0.9518)        ( 0.5998,   1.1354)
1.4474         0.0286      ( 1.3871,   1.5076)        ( 1.1861,   1.7086)
```

16.9 (a) Below (from Minitab). (b) H_0: $\beta_1 = 0$; this says that VO2 is not linearly related to HR. (c) If H_0 is true, F has a $F(1, 17)$ distribution; $F = 259.27$ has $P < 0.001$. (d) We found $t = 16.10$, and $t^2 = 259.21$. (e) $r^2 = $ SSM/SST $= 3.7619/4.0085 = 93.8\%$.

Output from Minitab

```
Analysis of Variance

SOURCE        DF        SS          MS          F          p
Regression     1      3.7619      3.7619     259.27     0.000
Error         17      0.2467      0.0145
Total         18      4.0085
```

16.10 The correlations are below. Of these, the correlation between GPA and IQ is largest in absolute value, so the relationship between them is closest to a straight line. About 40.2% of the variation in GPA is explained by the relationship with IQ.

IQ	0.634	C2	0.601
AGE	−0.389	C3	0.495
SEX	−0.097	C4	0.267
SC	0.542	C5	0.472
C1	0.441	C6	0.401

16.11 (a) With the given values, $\mu_{\text{GPA}} = \beta_0 + 9\beta_1 + 8\beta_2 + 7\beta_3$. (b) We estimate $\widehat{\text{GPA}} = 2.697$. Among all computer science students with the given high school grades, we expect the mean college GPA after three semesters to be about 2.7.

16.12 (a) With the given values, $\mu_{\text{GPA}} = \beta_0 + 6\beta_1 + 7\beta_2 + 8\beta_3$. (b) We estimate $\widehat{\text{GPA}} = 2.202$. Among all computer science students with the given high school grades, we expect the mean college GPA after three semesters to be about 2.2.

16.13 The critical value for df $= 220$ is $t^* \doteq 1.9708$. If using Table C, take $t^* = 1.984$. (a) $b_1 \pm t^*$ $SE_{b1} = 0.0986$ to 0.2385 (or 0.0982 to 0.2390). This coefficient gives the average increase in college GPA for each 1-point increase in high school math grade. (b) $b_3 \pm t^* SE_{b3} = -0.0312$ to 0.1214 (or

−0.0317 to 0.1219). This coefficient gives the average increase in college GPA for each 1-point increase in high school English grade.

16.14 The critical value for df = 221 is $t^* \doteq 1.9708$. If using the table, take $t^* = 1.984$. (a) $b_1 \pm t^*$ SE_{b1} = 0.1197 to 0.2456 (or 0.1193 to 0.2461). This coefficient gives the average increase in college GPA for each 1-point increase in high school math grade. (b) $b_2 \pm t^* SE_{b2}$ = −0.0078 to 0.1291 (or −0.0082 to 0.1296). This coefficient gives the average increase in college GPA for each 1-point increase in high school English grade. The coefficients (and standard errors) can change greatly when the model changes.

16.15 (a) $\widehat{GPA}$ = 0.590 + 0.169 HSM + 0.034 HSS + 0.045 HSE. (b) $s = \sqrt{MSE}$ = 0.69984. (c) $H_0: \beta_1 = \beta_2 = \beta_3 = 0$; H_a: at least one $\beta_j \neq 0$. In words, H_0 says that none of the high school grade variables are predictors of college GPA (in the form given in the model); H_a says that at least one of them is. (d) Under H_0, F has an F(3,220) distribution. Since P = 0.0001, we reject H_0. (e) The regression explains 20.46% of the variation in GPA.

16.16 (a) $\widehat{GPA}$ = 1.289 + 0.002283 SATM − 0.00002456 SATV. (b) $s = \sqrt{MSE}$ = 0.75770. (c) $H_0: \beta_1 = \beta_2 = 0$; H_a: at least one $\beta_j \neq 0$. In words, H_0 says that neither SAT score predicts college GPA (in the form given in the model); H_a says that at least one of them is a predictor. (d) Under H_0, F has an F(2, 221) distribution. Since P = 0.0007, we reject H_0. (e) The regression explains 6.34% of the variation in GPA.

16.17 The regression equation (given in the answer to Exercise 16.15 and Figure 16.6) is $\widehat{GPA}$ = 0.590 + 0.169 HSM + 0.034 HSS + 0.045 HSE. Among other things, we note that most of the residuals associated with low HS grades, and (not coincidentally) with low predicted GPAs, are "large" (positive, or just slightly negative). Also, using this model, the predicted GPAs are all between 1.33 and 3.07.

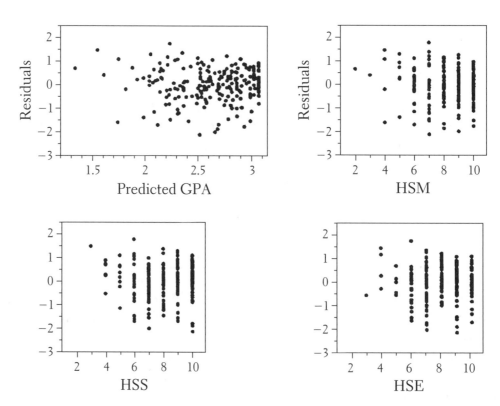

16.18 The regression equation (given in the answer to Exercise 16.16 and Figure 16.9) is $\widehat{GPA}$ = 1.289 + 0.002283 SATM − 0.00002456 SATV. The residual plots show no striking patterns, but one noticeable feature is the similarity between the predicted GPA and SATM plots—which results from the fact that the coefficient of SATV is so small that predicted GPA is *almost* a linear function of SATM alone.

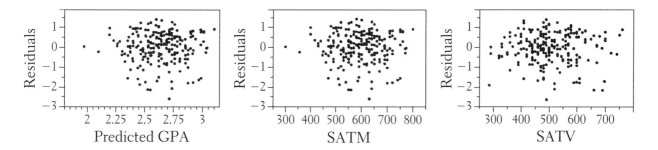

16.19 (a) $\widehat{GPA}$ = 0.666 + 0.193 HSM + 0.000610 SATM. (b) H_0: $\beta_1 = \beta_2 = 0$; H_a: at least one $\beta_j \neq 0$. In words, H_0 says that neither mathematics variable is a predictor of college GPA (in the form given in the model); H_a says that at least one of them is. The F statistic (with df = 2 and 221) is 26.63; this has $P < 0.0005$, so we reject H_0. Minitab output follows. (c) The critical value for df = 221 is $t^* \doteq 1.9708$. If using the table, take $t^* \doteq 1.984$. For the coefficient of HSM, $SE_{b1} = 0.03222$, so the interval is 0.1295 to 0.2565 (or 0.1291 to 0.2569). For the coefficient of SATM, $SE_{b2} = 0.0006112$, so the interval is -0.000594 to 0.001815 (or -0.000602 to 0.001823)—which contains 0. (d) HSM: $t = 5.99$, $P < 0.0005$. SATM: $t = 1.00$, $P = 0.319$. As the intervals indicated, the coefficient of SATM is not significantly different from 0. (e) $s = \sqrt{MSE} = 0.7028$. (f) The regression explains 19.4% of the variation in GPA.

Output from Minitab

```
The regression equation is
GPA = 0.666 + 0.193 HSM + 0.000610 SATM

Predictor          Coef        Stdev     t-ratio         p
Constant         0.6657       0.3435        1.94     0.054
HSM              0.19300      0.03222       5.99     0.000
SATM             0.0006105    0.0006112     1.00     0.319

s = 0.7028    R-sq = 19.4%    R-sq(adj) = 18.7%

Analysis of Variance

SOURCE        DF          SS         MS        F         p
Regression     2      26.303     13.151    26.63     0.000
Error        221     109.160      0.494
Total        223     135.463
```

16.20 The regression equation is $\widehat{GPA}$= 1.28 + 0.143 HSE + 0.000394 SATV, and R^2 = 8.6%. The regression is significant (F = 10.34, with df = 2 and 221); the t-tests reveal that the coefficient of SATV is not significantly different from 0 (t = 0.71, P = 0.481). For mathematics variables, we had R^2 = 19.4%—not overwhelmingly large, but considerably more than that for verbal variables.

Output from Minitab

```
The regression equation is
GPA = 1.28 + 0.143 HSE + 0.000394 SATV

Predictor          Coef        Stdev     t-ratio         p
Constant         1.2750       0.3474        3.67     0.000
HSE              0.14348      0.03428       4.19     0.000
SATV             0.0003942    0.0005582     0.71     0.481
```

```
s = 0.7487    R-sq = 8.6%    R-sq(adj) = 7.7%

Analysis of Variance

SOURCE          DF           SS          MS         F        p
Regression       2      11.5936      5.7968     10.34    0.000
Error          221     123.8692      0.5605
Total          223     135.4628
```

16.21 For males, regression gives $\widehat{GPA}= 0.582 + 0.155$ HSM $+ 0.0502$ HSS $+ 0.0445$ HSE, with $R^2 = 18.4\%$. The regression is significant ($F = 10.62$ with df $= 3$ and 141; $P < 0.0005$), but only the coefficient of HSM is significantly different from 0 (even the constant 0.582 has $t = 1.54$ and $P = 0.125$). Regression with HSM and HSS (excluding HSE since it has the largest P-value) gives the equation $\widehat{GPA}= 0.705 + 0.159$ HSM $+ 0.0738$ HSS, and $R^2 = 18.0\%$. The P-values for the constant and the coefficient of HSS are smaller (although the latter is still not significantly different from 0). One might also regress on HSM alone; this has $R^2 = 16.3\%$.

Minitab output for all three models follows. Residual plots (not shown) do not suggest problems with any of the models.

Comparing these results to the appropriate figures in the text, note that with all students, we excluded HSS (rather than HSE) in the second model.

Output from Minitab

```
The regression equation is
GPAm = 0.582 + 0.155 HSMm + 0.0502 HSSm + 0.0445 HSEm

Predictor       Coef      Stdev     t-ratio        p
Constant      0.5818     0.3767        1.54     0.125
HSMm          0.15502    0.04487       3.45     0.001
HSSm          0.05015    0.05070       0.99     0.324
HSEm          0.04446    0.05037       0.88     0.379

s = 0.7363    R-sq = 18.4%    R-sq (adj) = 16.7%


----------------- SECOND MODEL ------------------


The regression equation is
GPAm = 0.705 + 0.159 HSMm + 0.0738 HSSm

Predictor       Coef      Stdev     t-ratio        p
Constant      0.7053     0.3495        2.02     0.045
HSMm          0.15863    0.04465       3.55     0.001
HSSm          0.07383    0.04299       1.72     0.088

s = 0.7357    R-sq = 18.0%    R-sq (adj) = 16.8%


----------------- THIRD MODEL -----------------


The regression equation is
GPAm = 0.962 + 0.200 HSMm

Predictor       Coef      Stdev     t-ratio        p
Constant      0.9619     0.3181        3.02     0.003
HSMm          0.19987    0.03790       5.27     0.000

s = 0.7407    R-sq = 16.3%    R-sq (adj) = 15.7%
```

16.22 For females, regression gives $\widehat{GPA} = 0.648 + 0.205$ HSM $+ 0.0018$ HSS $+ 0.0324$ HSE, with $R^2 = 25.1\%$. In this equation, only the coefficient of HSM is significantly different from 0 (even the constant 0.648 has $t = 1.17$ and $P = 0.247$). Regression with HSM and HSE (excluding HSS since it has the largest P-value) gives the equation $\widehat{GPA} = 0.648 + 0.206$ HSM $+ 0.0333$ HSE, and $R^2 = 25.1\%$—but the P-values for the constant and coefficient of HSE have changed very little. With HSM alone, the regression equation is $\widehat{GPA} = 0.821 + 0.220$ HSM, R^2 decreases only slightly to 24.9%, and both the constant and coefficient are significantly different from 0.

Minitab output for all three models follows. Residual plots (not shown) do not suggest problems with any of the models.

Comparing the results to males, we see that both HSM and HSS were fairly useful for men, but HSM was sufficient for women—based on R^2, HSM alone does a better job for women than all three variables for men.

Output from Minitab

```
The regression equation is
GPAf = 0.648 + 0.205 HSMf + 0.0018 HSSf + 0.0324 HSEf

Predictor       Coef      Stdev    t-ratio         p
Constant      0.6484     0.5551       1.17     0.247
HSMf         0.20512    0.06134       3.34     0.001
HSSf         0.00178    0.05873       0.03     0.976
HSEf         0.03243    0.08270       0.39     0.696

s = 0.6431    R-sq = 25.1%    R-sq (adj) = 22.1%

---------------- SECOND MODEL -----------------


The regression equation is
GPAf = 0.648 + 0.206 HSMf + 0.0333 HSEf

Predictor       Coef      Stdev    t-ratio         p
Constant      0.6483     0.5514       1.18     0.243
HSMf         0.20596    0.05430       3.79     0.000
HSEf         0.03328    0.07732       0.43     0.668

s = 0.6389    R-sq = 25.1%    R-sq (adj) = 23.1%

------------------- THIRD MODEL -----------------


The regression equation is
GPAf = 0.821 + 0.220 HSMf

Predictor       Coef      Stdev    t-ratio         p
Constant      0.8213     0.3755       2.19     0.032
HSMf         0.21984    0.04347       5.06     0.000

s = 0.6355    R-sq = 24.9%    R-sq (adj) = 24.0%
```

16.23 (a) Regression gives $\widehat{GPA} = -2.83 + 0.0822$ IQ $+ 0.163$ C3, and $R^2 = 45.9\%$. For the coefficient of C3, $t = 2.83$, which has $P = 0.006$—significantly different from 0. C3 increases R^2 by $5.7\% = 45.9\% - 40.2\%$.

(b) Regression now gives $\widehat{GPA} = -3.49 + 0.0761$ IQ $+ 0.0670$ C3 $+ 0.0369$ SC, and $R^2 = 47.5\%$. For the coefficient of C3, $t = 0.78$, which has $P = 0.436$—*not* significantly different from 0. When self-concept (SC) is included in the model, C3 adds little. (If we regress on IQ and SC, $R^2 = 47.1\%$.)

(c) The values change because coefficients are quite sensitive to changes in the model, especially when the explanatory variables are highly correlated (the correlation between SC and C3 is about 0.80). In this case, the predictive information of SC and C3 overlap, so that the two of them together add little more than either one separately (with IQ).

Output from Minitab

```
The regression equation is
GPA = -2.83 + 0.0822 IQ + 0.163 C3

Predictor        Coef       Stdev     t-ratio        p
Constant       -2.829       1.507       -1.88    0.064
IQ            0.08220     0.01508        5.45    0.000
C3           0.16289     0.05752        2.83    0.006

s = 1.564    R-sq = 45.9%    R-sq (adj) = 44.5%
```

---------------- SECOND MODEL ----------------

```
The regression equation is
GPA = -3.49 + 0.0761 IQ + 0.0670 C3 + 0.0369 SC

Predictor        Coef       Stdev     t-ratio        p
Constant       -3.491       1.558       -2.24    0.028
IQ            0.07612     0.01549        4.91    0.000
C3           0.06701     0.08558        0.78    0.436
SC           0.03691     0.02456        1.50    0.137

s = 1.551    R-sq = 47.5%    R-sq (adj) = 45.4%
```

16.24 (a) Regression gives $\widehat{\text{GPA}} = -4.94 + 0.0815\,\text{IQ} + 0.183\,\text{C1} + 0.142\,\text{C5}$, $R^2 = 52.5\%$, and $s = 1.475$. With the given values of IQ, C1, and C5, $\widehat{\text{GPA}} = 7.457$.

(b) GPA would increase by about 0.0815 per IQ point (the coefficient of IQ). $\text{SE}_{b1} = 0.01367$; with df = 74, $t^* = 1.9926$ (or use $t^* = 2.000$ from the table). Interval: 0.0543 to 0.1087 (or 0.0542 to 0.1088).

(c) The residual plots are below. The residual for OBS = 55 stands out as being extraordinarily low; this student had the lowest GPA and, at 15 years old, was the oldest.

(d) Regression now gives $\widehat{\text{GPA}} = -4.68 + 0.0805\,\text{IQ} + 0.197\,\text{C1} + 0.109\,\text{C5}$ $R^2 = 57.4\%$, and $s = 1.303$. With the given values of IQ, C1, and C5, $\widehat{\text{GPA}} = 7.534$. Removing this observation did not greatly change the model or the prediction, although the coefficient of C5 is not quite significant under the new regression (see Minitab output on the next page).

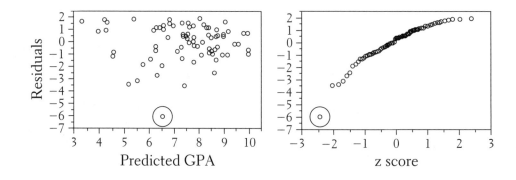

Output from Minitab

```
The regression equation is
GPA = -4.94 + 0.0815 IQ + 0.183 C1 + 0.142 C5

Predictor        Coef       Stdev     t-ratio         p
Constant       -4.937       1.491       -3.31     0.001
IQ            0.08145     0.01367        5.96     0.000
C1            0.18308     0.06475        2.83     0.006
C5            0.14205     0.06663        2.13     0.036

s = 1.475    R-sq = 52.5%    R-sq (adj) = 50.6%
```

---------------------------- Without OBS 55 ----------------------------

```
The regression equation is
GPA = -4.68 + 0.0805 IQ + 0.197 C1 + 0.109 C5

Predictor        Coef       Stdev     t-ratio         p
Constant       -4.678       1.318       -3.55     0.001
IQ            0.08050     0.01207        6.67     0.000
C1            0.19707     0.05724        3.44     0.001
C5            0.10950     0.05923        1.85     0.069

s = 1.303    R-sq = 57.4%    R-sq (adj) = 55.7%
```

16.25 The regression equation is $\widehat{Taste}= -26.9 + 3.80\,Acetic + 5.15\,H2S$. The model explains 58.2% of the variation in Taste. The t-value for the coefficient of Acetic is 0.84 ($P = 0.406$), indicating that it does not add significantly to the model when H2S is used, because Acetic and H2S are correlated (in fact, $r = 0.618$ for these two variables). This model does a better job than any of the three simple linear regression models, but it is not much better than the model with H2S alone (which explained 57.1% of the variation in Taste)—as we might expect from the t-test result.

16.26 The regression equation is $\widehat{Taste}= -27.6 + 3.95\,H2S + 19.9\,Lactic$. The model explains 65.2% of the variation in Taste, which is higher than for the two simple linear regressions. Both coefficients are significantly different from 0 ($P = 0.002$ for H2S, and $P = 0.019$ for Lactic).

16.27 The regression equation is $\widehat{Taste}= -28.9 + 0.33\,Acetic + 3.91\,H2S + 19.7\,Lactic$. The model explains 65.2% of the variation in Taste (the same as for the model with only H2S and Lactic). Residuals of this regression are positively associated with Taste, but they appear to be normally distributed and show no patterns in scatterplots with other variables.

The coefficient of Acetic is not significantly different from 0 ($P = 0.942$); there is no gain in adding Acetic to the model with H2S and Lactic. It appears that the best model is the H2S/Lactic model of Exercise 16.26.

17

Logistic Regression

17.1 (a) For the high blood pressure group, $\hat{p} = \frac{55}{3338} \doteq 0.01648$, giving odds $\frac{\hat{p}}{1-\hat{p}} = \frac{55}{3283} \doteq 0.01675$, or about 1 to 60. (If students give odds in the form "a to b," their choices of a and b might be different.)

(b) For the low blood pressure group, $\hat{p} = \frac{21}{2676} \doteq 0.00785$, giving odds $\frac{\hat{p}}{1-\hat{p}} = \frac{21}{2655} \doteq 0.00791$, or about 1 to 126 (or 125).

(c) The odds ratio is about 2.118. Odds of death from cardiovascular disease are about 2.1 times greater in the high blood pressure group.

17.2 (a) For female references, $\hat{p} = \frac{48}{60} = 0.8$, giving odds $\frac{\hat{p}}{1-\hat{p}} = \frac{48}{12} = 4$ ("4 to 1"). (b) For male references, $\hat{p} = \frac{52}{132} = 0.\overline{39}$, giving odds $\frac{\hat{p}}{1-\hat{p}} = \frac{52}{80} = 0.65$ ("13 to 20"). (c) The odds ratio is about 6.154. The odds of a juvenile reference are more than six times greater for females.

17.3 (a) See below. The curve generated here is the same as the curve in Figure 17.1.

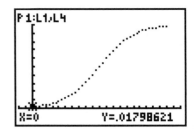

(b) See below.

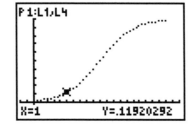
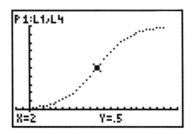

(c) Solving $\ln\left(\dfrac{p}{1-p}\right) = -4 + 2(4) = 4$ for p, we obtain

242

$$\frac{p}{1-p} = e^4 = 54.59815 \qquad p = 54.59815 - 54.59815p$$
$$55.59815p = 54.59815 \qquad p = 0.982$$

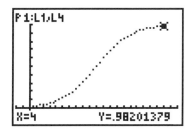

The plot above verifies that $p = 0.982$ when $x = 4$.

17.4 For the model $\log\left(\frac{p}{1-p}\right) = \beta_0 + \beta_1 x$, we obtain the fitted model $\log(\text{ODDS}) = b_0 + b_1 x = -7.2789 + 0.9399x$. (Here p is the probability that the cheese is acceptable, and x is the value of H2S.) We have $b_1 = 0.9399$ and $\text{SE}_{b_1} = 0.3443$, so we estimate that the odds ratio increases by a factor of $e^{b_1} \doteq 2.56$ for every unit increase in H2S. For testing $\beta_1 = 0$, we find $X^2 = 7.45$ ($P = 0.0063$), so we conclude that $\beta_1 \neq 0$. We are 95% confident that β_1 is in the interval $b_1 \pm 1.96\text{SE}_{b_1} = 0.2651$ to 1.6147; exponentiating this tells us that the odds ratio increases by a factor between 1.3035 and 5.0265 (with 95% confidence) for each unit increase in H2S.

17.5 For the model $\log\left(\frac{p}{1-p}\right) = \beta_0 + \beta_1 x$, we obtain the fitted model $\log(\text{ODDS}) = b_0 + b_1 x = -10.7799 + 6.3319x$. (Here p is the probability that the cheese is acceptable, and x is the value of Lactic.) We have $b_1 = 6.3319$ and $\text{SE}_{b_1} = 2.4532$, so we estimate that the odds ratio increases by a factor of $e^{b_1} \doteq 562.22$ for every unit increase in Lactic. For testing $\beta_1 = 0$, we find $X^2 = 6.66$ ($P = 0.0098$), so we conclude that $\beta_1 \neq 0$. We are 95% confident that β_1 is in the interval $b_1 \pm 1.96\text{SE}_{b_1} = 1.5236$ to 11.1402; exponentiating this tells us that the odds ratio increases by a factor between 4.5889 and about 68.884 (with 95% confidence) for each unit increase in Lactic.

17.6 (a) Data below left, plot (with overlaid least squares line) below right. The plot does resemble the plot of Figure 17.3.

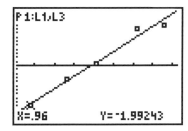

(b) $\ln(\text{ODDS}) = -4.89 + 3.10 \ln(\text{CONC})$. (c) $r^2 = 0.9789$. 97.89% of the variation in $\ln(\text{ODDS})$ can be explained by linear regression on $\ln(\text{CONC})$.

17.7 The output of the Logistic command is given below left, the plot (with overlaid curve) below right. The curve is almost identical to that of Figure 17.4.

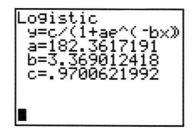 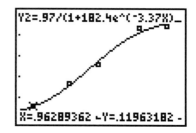

17.8 (a) Find $b_1 \pm z^* \, SE_{b_1}$, using either $z^* = 2$ or $z^* = 1.96$. These give 0.2349 to 1.2661, or 0.2452 to 1.2558, respectively. (b) $X^2 = \left(\frac{0.7505}{0.2578}\right)^2 \doteq 8.47$. This gives a P-value between 0.0025 and 0.005. (c) We have strong evidence that there is a real (significant) difference in risk between the two groups.

17.9 (a) Find $b_1 \pm z^* \, SE_{b_1}$, using either $z^* = 2$ or $z^* = 1.96$. These give 1.0799 to 2.5543, or 1.0946 to 2.5396, respectively. (b) $X^2 = \left(\frac{0.8171}{0.3686}\right)^2 \doteq 24.3023$. This gives $P < 0.0005$. (c) We have strong evidence that there is a real (significant) difference in juvenile references between male and female references.

17.10 The seven models are summarized below. The P-value in the right column is for the null hypothesis that all slopes equal 0 (i.e., the significance of the regression); all are significant.

For the three new models (those with two predictors), all have only one coefficient significantly different from 0 (in the last case, arguably neither coefficient is nonzero). The standard errors are given in parentheses below each coefficient; the six respective P-values are 0.4276, 0.0238; 0.3094, 0.0355; 0.0567, 0.1449.

In summary, we might conclude that H2S has the greatest effect: it had the smallest P-value among the three single-predictor models, and in the three multiple logistic regression models in which it was used, it had the minimum P-value. (It was the closest to being significant in the last two models in the table below.)

Fitted Model	P
log(ODDS) = −13.71 + 2.249 Acetic	0.0285
log(ODDS) = −7.279 + 0.9399 H2S	0.0063
log(ODDS) = −10.78 + 6.332 Lactic	0.0098
log(ODDS) = −12.85 + 1.096 Acetic + 0.8303 H2S (1.382) (0.3673)	0.0008
log(ODDS) = −16.56 + 1.309 Acetic + 5.257 Lactic (1.288) (2.500)	0.0016
log(ODDS) = −11.72 + 0.7346 H2S + 3.777 Lactic (0.3866) (2.596)	0.0003
log(ODDS) = −14.26 + 0.584 Acetic + 0.6849 H2S + 3.468 Lactic	0.0010

17.11 (a) The estimated odds ratio is $e^{b_1} \doteq 2.118$ (as we found in Exercise 17.1). Exponentiating the intervals for β_1 in Exercise 17.8 (a) gives odds-ratio intervals from about 1.26 to 3.55 ($z^* = 2$), or 1.28 to 3.51 ($z^* = 1.96$).

(b) We are 95% confident that the odds of death from cardiovascular disease are about 1.3 to 3.5 times greater in the high blood pressure group.

17.12 (c) The estimated odds ratio is $e^{b_1} \doteq 6.154$ (as we found in Exercise 17.2). Exponentiating the intervals for β_1 in Exercise 17.9 (a) gives odds-ratio intervals from about 2.94 to 12.86 ($z^* = 2$), or 2.99 to 12.67 ($z^* = 1.96$). (b) We are 95% confident that the odds of a juvenile reference are about 3 to 13 times greater among females.

17.13 (a) The model is $\log\left(\frac{p_i}{1 - p_i}\right) = \beta_0 + \beta_1 x_i$, where $x_i = 1$ if the ith person is over 40, and 0 if he/she is under 40.

(b) p_i is the probability that the ith person is terminated; this model assumes that the probability of termination depends on age (over/under 40). In this case, that seems to have been the case, but we might expect that other factors were taken into consideration.

(c) The estimated odds ratio is $e^{b_1} \doteq 3.859$. (Of course, we can also get this from $\frac{41/765}{7/504}$.) We can also find, e.g., a 95% confidence interval for b_1: $b_1 \pm 1.96\ \mathrm{SE}_{b_1} = 0.5409$ to 2.1599. Exponentiating this translates to a 95% confidence interval for the odds: 1.7176 to 8.6701. The odds of being terminated are 1.7 to 8.7 times greater for those over 40.

(d) Use a multiple logistic regression model, e.g., $\log\left(\frac{p_i}{1 - p_i}\right) = \beta_0 + \beta_1 x_i + \beta_2 y_i$.

17.14 We show the steps for doing this by hand: if software is available, the results should be the same. The model is $\log \log\left(\frac{p}{1 - p}\right) = \beta_0 + \beta_1 x$. We make the arbitrary choice to take x to be the indicator variable for "male"—i.e., $x = 1$ for men, 0 for women. (We could also choose to have $x = 1$ for women, and 0 for men.) Then

$$\log\left(\frac{p_m}{1 - p_m}\right) = \beta_0 + \beta_1, \text{ and } \log\left(\frac{pf}{1 - pf}\right) = \beta_0.$$

With the given data, we estimate

$$\log\left(\frac{\hat{p}_m}{1 - \hat{p}_m}\right) = \log\left(\frac{515}{1005}\right) \doteq -0.6686 = b_0 + b_1 \quad \text{and}$$

$$\log\left(\frac{\hat{pf}}{1 - \hat{pf}}\right) = \log\left(\frac{27}{164}\right) \doteq -1.8040 - b_0$$

so we find that $b_0 = -1.8040$ and $b_1 = 1.1354$. This gives an odds ratio of about $e^{b_1} \doteq 3.11$; we estimate that the odds for a male testing positive are about three times those for a female.

With software, we find $\mathrm{SE}_{b_1} \doteq 0.2146$ and $X^2 = 27.98$ ($P < 0.0001$). The logistic regression is significant (i.e., we conclude that $\beta_1 \neq 0$). A 95% confidence interval for β_1 is $b_1 \pm 1.96\mathrm{SE}_{b_1} = 0.7148$ to 1.5561, so we are 95% confident that the odds ratio is between about 2.04 and 4.74.

17.15 Portions of SAS and GLMStat output are given below. (a) The X^2 statistic for testing this hypothesis is 33.65 (df = 3), which has $P = 0.0001$. We conclude that at least one coefficient is not 0. (b) The model is $\log(\mathrm{ODDS}) = -6.053 + 0.3710\mathrm{HSM} + 0.2489\mathrm{HSS} + 0.03605\mathrm{HSE}$. The standard errors of the three coefficients are 0.1302, 0.1275, and 0.1253, giving respective 95% confidence intervals 0.1158 to 0.6262, −0.0010 to 0.4988, and −0.2095 to 0.2816. (c) Only the coefficient of HSM is significantly different from 0, though HSS may also be useful. (Only HSM was useful in the multiple linear regression model of GPA on high school grades.)

Output from SAS:

```
                               Intercept
                  Intercept      and
Criterion          Only       Covariates      Chi-Square for Covariates
AIC               297.340       269.691               .
SC                300.751       283.338               .
-2 LOG L          295.340       261.691         33.648 with 3 DF (p = 0.0001)
Score                .             .            29.672 with 3 DF (p = 0.0001)
```

Analysis of Maximum Likelihood Estimates

Variable	DF	Parameter Estimate	Standard Error	Wald Chi-Square	Pr > Chi Square	Standardized Estimate
INTERCPT	1	−6.0528	1.1562	27.4050	0.0001	.
HSM	1	0.3710	0.1302	8.1155	0.0044	0.335169
HSS	1	0.2489	0.1275	3.8100	0.0509	0.233265
HSE	1	0.0361	0.1253	0.0828	0.7736	0.029971

Output from GLMStat:

	estimate	se(ast)	z ratio	Prob>|z|
1 Constant	−6.053	1.156	−5.236	<0.0001
2 HSM	0.3710	0.1302	2.849	0.0044
3 HSS	0.2489	0.1275	1.952	0.0509
4 HSE	3.605e−2	0.1253	0.2877	0.7736

17.16 Portions of SAS and GLMStat output are given below. (a) The X^2 statistic for testing this hypothesis is 14.2 (df = 2), which has $P = 0.0008$. We conclude that at least one coefficient is not 0. (b) The model is log(ODDS) = −4.543 + 0.003690 SATM + 0.003527 SATV. The standard errors of the two coefficients are 0.001913 and 0.001751, giving respective 95% confidence intervals −0.000059 to 0.007439, and 0.000095 to 0.006959. (The first coefficient has a P-value of 0.0537, and the second has $P = 0.0440$.) (c) We (barely) cannot reject $\beta_{SATM} = 0$—though since 0 is just in the confidence interval, we are reluctant to discard SATM. Meanwhile, we conclude that $\beta_{SATV} \neq 0$. (By contrast, with multiple linear regression of GPA on SAT scores, we found SATM useful, but not SATV.)

Output from SAS:

```
                               Intercept
                  Intercept      and
Criterion          Only       Covariates      Chi-Square for Covariates
AIC               297.340       287.119               .
SC                300.751       297.354               .
-2 LOG L          295.340       281.119         14.220 with 2 DF (p = 0.0008)
Score                .             .            13.710 with 2 DF (p = 0.0011)
```

Analysis of Maximum Likelihood Estimates

Variable	DF	Parameter Estimate	Standard Error	Wald Chi-Square	Pr > Chi-Square	Standardized Estimate
INTERCPT	1	−4.5429	1.1618	15.2909	0.0001	
SATM	1	0.00369	0.00191	3.7183	0.0538	0.175778
SATV	1	0.00353	0.00175	4.0535	0.0441	0.180087

Output from GLMStat:

		estimate	se (est)	z ratio	Prob>|z|
1	Constant	−4.543	1.161	−3.915	<0.0001
2	SATM	3.690e−3	1.913e−3	1.929	0.0537
3	SATV	3.527e−3	1.751e−3	2.014	0.0440

17.17 The coefficients and standard errors for the fitted model are below. (a) The X^2 statistic for testing this hypothesis is 23.0 (df = 3); since $P < 0.0001$, we reject H_0 and conclude that high school grades add a significant amount to the model with SAT scores. (b) The X^2 statistic for testing this hypothesis is 3.6 (df = 2); since $P = 0.1653$, we cannot reject H_0. SAT scores do not add significantly to the model with high school grades. (c) For modeling the odds of HIGPA, high school grades (specifically HSM, and to a lesser extent HSS) are useful, while SAT scores are not.

Output from SAS:

Analysis of Maximum Likelihood Estimates

Variable	DF	Parameter Estimate	Standard Error	Wald Chi-Square	Pr> Chi-Square	Standardized Estimate
INTERCPT	1	−7.3732	1.4768	24.9257	0.0001	
HSM	1	0.3427	0.1419	5.8344	0.0157	0.309668
HSS	1	0.2249	0.1286	3.0548	0.0805	0.210704
HSE	1	0.0190	0.1289	0.0217	0.8829	0.015784
SATM	1	0.000717	0.00220	0.1059	0.7448	0.034134
SATV	1	0.00289	0.00191	2.2796	0.1311	0.147566

Output from GLMStat:

		estimate	se(est)	z ratio	Prob>|z|
1	Constant	−7.373	1.477	−4.994	<0.0001
2	SATM	7.166e−4	2.201e−3	0.3255	0.7448
3	SATV	2.890e−3	1.914e−3	1.510	0.1311
4	HSM	0.3427	0.1419	2.416	0.0157
5	HSS	0.2249	0.1286	1.748	0.0805
6	HSE	1.899e−2	0.1289	0.1473	0.8829

17.18 (a) The fitted model is log(ODDS) = −0.6124 + 0.0609 Gender; the coefficient of gender is not significantly different from 0 ($SE_{b_{Gender}} = 0.2889$, $P = 0.8331$).
(b) Now log(ODDS) = −5.214 + 0.3028 Gender + 0.004191 SATM + 0.003447 SATV. In this model, gender is still not significant ($P = 0.3296$).
(c) Gender is not useful for modeling the odds of HIGPA.

Output from GLMStat:

		estimate	se (est)	z ratio	Prob>|z|
1	Constant	−5.214	1.362	−3.828	0.0001
2	Gender	0.3028	0.3105	0.9750	0.3296
3	SATM	4.191e−3	1.987e−3	2.109	0.0349
4	SATV	3.447e−3	1.760e−3	1.958	0.0502

17.19 (a) The fitted model is log(ODDS) = 3.4761 + 0.4157x, x = 0 for Hospital A and 1 for Hospital B. With $b_1 \doteq 0.4157$ and $SE_{b_1} \doteq 0.2831$, we find that $X^2 = 2.16$ ($P = 0.1420$), so we do not have evidence to suggest that β_1 is not 0. A 95% confidence interval for β_1 is -0.1392 to 0.9706 (this interval includes 0). We estimate the odds ratio to be $e^{b_1} \doteq 1.52$, with confidence interval 0.87 to 2.64 (this includes 1, since β_1 might be 0). (b) The fitted model is log(ODDS) = -6.930 + 1.009 Hospital $-$ 0.09132 Condition; as before, use 0 for Hospital A and 1 for Hospital B, and 1 for good condition, and 0 for poor. Now we estimate the odds ratio to be $e^{b_1} \doteq 2.74$, with confidence interval 0.30 to 25.12. (c) In neither case is the effect significant; Simpson's paradox is seen in the increased width of the interval from part (a) to part (b).

Output from GLMStat:

		estimate	se(est)	z ratio	Prob>|z|
1	Constant	-6.930	0.7693	-9.009	<0.0001
2	Hosp	1.009	1.130	0.8928	0.3720
3	Cond	$-9.132e-2$	1.130	8.080e-2	0.9356